T0245306

Applied Operational Excellence for the Oil, Gas, and Process Industries

Applied Operational Excellence for the Oil, Gas, and Process Industries

Dennis P. Nolan
Eric T. Anderson

AMSTERDAM • BOSTON • HEIDELBERG • LONDON
NEW YORK • OXFORD • PARIS • SAN DIEGO
SAN FRANCISCO • SINGAPORE • SYDNEY • TOKYO
Gulf Professional Publishing is an imprint of Elsevier

Gulf Professional Publishing is an imprint of Elsevier
225 Wyman Street, Waltham, MA 02451, USA
The Boulevard, Langford Lane, Kidlington, Oxford, OX5 1GB, UK

Copyright © 2015 Elsevier Inc. All rights reserved.

No part of this publication may be reproduced or transmitted in any form or by any means, electronic or mechanical, including photocopying, recording, or any information storage and retrieval system, without permission in writing from the publisher. Details on how to seek permission, further information about the Publisher's permissions policies and our arrangements with organizations such as the Copyright Clearance Center and the Copyright Licensing Agency, can be found at our website: www.elsevier.com/permissions.

This book and the individual contributions contained in it are protected under copyright by the Publisher (other than as may be noted herein).

Notices
Knowledge and best practice in this field are constantly changing. As new research and experience broaden our understanding, changes in research methods, professional practices, or medical treatment may become necessary.

Practitioners and researchers must always rely on their own experience and knowledge in evaluating and using any information, methods, compounds, or experiments described herein. In using such information or methods they should be mindful of their own safety and the safety of others, including parties for whom they have a professional responsibility.

To the fullest extent of the law, neither the Publisher nor the authors, contributors, or editors, assume any liability for any injury and/or damage to persons or property as a matter of products liability, negligence or otherwise, or from any use or operation of any methods, products, instructions, or ideas contained in the material herein.

ISBN: 978-0-12-802788-2

Library of Congress Cataloging-in-Publication Data
A catalogue record for this book is available from the Library of Congress

British Library Cataloguing-in-Publication Data
A catalogue record for this book is available from the British Library

For Information on all Gulf Professional Publishing publications
visit our website at http://store.elsevier.com/

Working together
to grow libraries in
developing countries

www.elsevier.com • www.bookaid.org

In all things success depends on previous preparation, and without such preparation there is sure to be failure.
Confucius (The Doctrine of the Mean)

Contents

About the Authors

Eric T. Anderson is currently a Loss Prevention specialist at Saudi Aramco. Before joining the Saudi Aramco Loss Prevention Department staff in 1998, he had numerous achievements to his credit throughout many years of safety, health, environmental and loss prevention management experience in the oil, gas, and petrochemical industry. Prior to joining Saudi Aramco he held a number of progressively responsible technical and managerial roles within the Kerr McGee Corporation, Imperial Chemical Industries (ICI) Group's North America based chemical operations, as well both Health, Environment and Safety (HES) organizations of Unocal and Chevron Corporation (pre/post-merger). He had a lead role in developing the Saudi Aramco Safety Management System (SMS) and also worked to help influence development efforts of the company's Operational Excellence Management System (OEMS). He has been elected to serve dutifully as an elected independent director for multiple terms with a Saudi Aramco employee self-directed group, the Arabian Automobile Association (AAA).

Eric has led numerous high profile incident investigations within the industry including analysis of a well control incident resulting in loss of an offshore platform in Southeast Asia, forensic investigation into causes of a California refinery incident involving loss of containment and off-site release with public health consequences (prompting a billion dollar class action lawsuit), a large fire and explosion within a process plant requiring the management of company resources in a comprehensive joint investigation that involved close coordination with Federal investigators from the US Department of Labor (OSHA) and the US Department of Defense. He has briefed members of the Board of Directors on technical issues related to several sensitive company incident investigations and advised on corrective actions. He has been the recipient of numerous recognition, innovation and excellence awards, including an ICI president's recognition and Chevron president's award as well as being profiled in company newsletters.

Eric earned his Bachelor of Science degree at the University of Wisconsin, Superior in 1983, his Master's degree from the University of Minnesota, Duluth in 1985 and he did post graduate work at Stanford University, Graduate School of Business where he completed the executive education program in Finance Management in 1997. Eric is currently an active professional member of the American Society of Safety Engineers (ASSE), a standards setting committee member and long standing professional member of the Society of Automotive Engineers (SAE), a Saudi Aramco representative to the Saudi Arabian Standards Organization (SASO), as well as serving as a former contributor and member of several American Petroleum Institute (API) technical standards committees. Over the course of his career, he has supported, audited, and inspected petroleum industry facilities across operations in the upstream, downstream, and midstream segments, facilities, and operations associated with production of natural gas, as well as onshore, offshore, and deepwater pipeline facilities. He has had assignments supporting both commercial and military explosives manufacturing facilities, surface, and underground mining operations, petroleum industry marketing facilities, and large scale geothermal operations under various operating conditions around the world. His activity in the chemical industry has included support and oversight for specialized facilities responsible for producing solid rocket fuel oxidizer used in NASA's Space Shuttle program, airbag initiator systems for the automotive industry, as well as components manufactured for "classified" missile defense systems.

Dr Dennis P. Nolan has had a long career devoted to risk engineering, fire protection engineering, loss prevention engineering, and systems safety engineering. He holds a Doctor of Philosophy degree in Business Administration from Berne University, Master of Science degree in Systems Management from Florida Institute of Technology and a Bachelor of Science Degree in Fire Protection Engineering from the University of Maryland. He is a US registered professional engineer (P.E.) in fire protection engineering in the state of California.

He is currently on the Executive Management staff of Saudi Aramco, located in Dhahran, Saudi Arabia, as a Loss Prevention Consultant/Chief Fire Prevention Engineer. He covers some of the largest oil and gas facilities in the world. The magnitude of the risks, worldwide sensitivity, and foreign location make this one the highly critical fire risk operations in the world. He has also been associated with Boeing, Lockheed, Marathon Oil Company, and Occidental Petroleum Corporation in various fire protection engineering, risk analysis, and safety roles in several locations in the United States and overseas. As part of his career, he has examined oil production, refining, and marketing facilities under severe conditions and in various unique worldwide locations, including Africa, Asia, Europe, the Middle East, Russia, and North, and South America. His activity in the aerospace field has included engineering support for the NASA Space Shuttle launch and landing facilities at Kennedy Space Center (and for those undertaken at Vandenburg Air Force Base, California) and "classified" national defense systems.

Dr Nolan has received numerous safety awards and is a member of the ASSE. He is the author of many technical papers and professional articles in various international fire protection and safety publications. He has written five other books, which include: *Safety and Security Review for the Process Industries, Application of HAZOP, PHA, What-IF and SVA Reviews* (1st, 2nd, 3rd, and 4th Editions), Handbook of Fire and Explosion Protection Engineering Principles for Oil, Gas, Chemical and Related Facilities (1st, 2nd, and 3rd Editions), *Fire Fighting Pumping Systems at Industrial Facilities* (1st & 2nd Editions), *Encyclopedia of Fire Protection* (1st & 2nd Editions), and *Loss Prevention and Safety Control Terms and Definitions*. Dr. Nolan has also been listed for many years in "Who's Who in California", Who's Who in the West", Who's Who in the World" and Who's Who in Science and Engineering" publications. He was also listed in "Outstanding Individuals of the twentieth Century" (2001) and "Living Legends" (2004), published by the International Biographical Center, Cambridge, England.

Preface

The material in this reference book is intended to provide guidance, insight, and background on a multitude of business and organizational issues relating to operational excellence that may help line management in many business segments. There is discussion of examples that have greatly influenced oil, gas, and related process industries and the evolution of operational excellence, but the issues are relevant in many other industries as well. The information provided can be used as a practical reference to prepare an organization's Operational Excellence management systems and also continually improve upon it.

To help facilitate its preparation we have reviewed many process organizations and industry guidelines on Operational Excellence, System Management Systems, and Quality programs. Many of these organizations have established proprietary processes, and although unique, and useful to their own operations, have not been highlighted since they would not have wide appeal. However, we have tried to capture and highlight the most relevant and important features of all of these programs in order to provide some background and understanding of what influences industry's efforts in achieving Operational Excellence. There are some areas in the book where we compare major organizations programs against each other. We have also tried to include the most useful and practical knowledge and applications that can be readily adapted for your use.

Introduction

Background of Safety, Health&Environment (SHE) and Operational Excellence Programs within the Oil, Gas, and Petrochemical Industries

The birth of the modern oil and gas industry is widely accepted to have been marked by the discovery at the "Lucas 1" well on January 10, 1901 at the famed Spindletop formation near Beaumont Texas, After drilling to depths, and stopping to pull the drill string out to change some equipment, the crew began the work to lower it all back in the hole. After lowering it back into the open hole to a depth of about 700 feet, more than a half day since they had made any forward progress with actual "drilling" activity, mud began bubbling back up the hole. Moments later, the drill pipe shot up out of the ground with tremendous force, and then, in what proved to be a dramatic pause, nothing seemed to happen.

After assessing the situation for a short while, a frustrated and confused drilling crew began cleaning up the mess and tried to determine if anything could be salvaged. All of a sudden and without warning, a noise that sounded like a cannon shot came from the hole, and mud began shooting out of the ground straight up into the air like a geyser. Within seconds, natural gas and oil followed. The oil "gusher," greenish-black in color, shot straight up into the air to a height that enveloped the drilling derrick, rising to a height of more than 150 feet before falling back to the surface. To put the magnitude of this event into context, this was more oil than had ever been seen anywhere in the entire world, it flowed at an initial rate of nearly 100,000 barrels per day, more than all of the other producing wells in the United States—*combined*. It was also an event that clearly demonstrated a genuine need for operational excellence; uncontrolled releases of material are *neither safe, nor profitable*, although maintaining well control and the means for regaining control of a blowout situation today is part of nearly every well prepared driller's emergency response plan.

With such an auspicious start it is no surprise that over the course of the industry's first 100 years, modern the oil, gas, and petrochemical industry has generally been perceived as a messy, sloppy industry. As the industry grew and took shape, it attracted men willing to take large risks to earn their fortunes. They would become stereotyped as rough and tough characters with outsized egos with more concern for profits than for safety or preserving the environment.

Oil Industry Rough Neck

Indeed, the petroleum industry's "rough neck" has been glamorized for years in media, music, and the movies. There are a great many men who have held this job title throughout the industry's history, and many revere it as a badge of honor. In the popular television show "Dallas" the main fictional character J.R. Ewing (played by Larry Hagman) was portrayed as a flamboyant oilman running a large Texas based oil firm enjoying an extraordinarily opulent lifestyle. In the show, he is depicted as a covetous, egocentric, manipulative, and amoral oil baron with psychopathic tendencies, constantly plotting subterfuges to plunder his foes and their wealth. Not surprisingly, reflecting upon these images and characterizations engrained in pop culture helped shape public perception of many of the industry's participants. And as with many things in life, it stands to reason that at least some of the stereotypes established for the industry were somewhat deserved.

With the new Industrial Age, demand for oil, gas, and petrochemical products continued to increase at a dizzying pace. Analyzing the frequency and magnitude of losses the industry experienced as it grew indicates that incidents involving fire, explosion, vapor cloud explosions, blowouts, structural failures, mechanical damage, and weather related events were occurring with ever increasing frequency. It became clear that industry loss incidents were not only increasing in both frequency but also in the magnitude of capital losses. For example, reflecting upon Marsh & McLennan loss data as well as the Lloyds Weekly Casualty Reports (LWCRs), Crook observed that until 1989 losses were doubling with each decade. Similarly, a study by V.C. Marshall (1975) showed that the number of vapor cloud explosions worldwide was also staging an apparent increasing trend.

Some losses garnered significant media attention and were at the forefront of public discussion and played a large part in shaping public perceptions; for others very little public interest was expressed.

The loss of the Deepwater Horizon drilling rig in the Gulf of Mexico made international headlines. Eleven fatalities and an uncontrolled hydrocarbon release lasting 87 days led to the total costs amounting to tens of billions of US dollars. The property damage value for this loss was just the tip of the iceberg. By way of contrast, losses resulting from a Floating Production, Storage and Offloading vessel (FSPO) moving off-station under heavy weather in the North Sea resulted in a property damage loss very much comparable to that of the Deepwater Horizon loss. However, the North Sea incident resulted in no loss of life and had very little environmental impact. In this case, the mainstream media showed little interest and it was reported primarily as a successful personnel rescue operation.

The industry's trend towards deployment of a rigorous Operational Excellence approach through many organizations within the process industry (i.e., oil, gas, petrochemical and chemical), has originated in response to a very competitive business landscape as well as a general direction or tendency leading to an increasingly more litigious society (in terms of both civil and criminal litigation) and the ever increasing regulatory governance structure throughout the industrialized world. High profile industry loss incidents have placed the industry squarely in the spotlight, prompting public inquires, grassroots advocacy, and shareholder activists all pushing corporate boards to provide greater transparency as well as ethical and socially responsible behaviors in the areas of health, environment, and safety.

Process Safety Management (PSM)/Risk Management Plan (RMP)

Major process incidents in the late 1980s (Texas City, TX (1987), Piper Alpha (1988), Bhopal, India (1989), Pasadena, TX (1989), etc.), resulted in considerable additional regulatory rulemaking in numerous countries that served to put additional pressure on management teams of companies both large and small. In the early 1980s the so-called "employee right to know" rule involving hazard communication was put in place to assure that companies have informed their employees of the hazards of chemicals they used inside the facility's fence line. The idea for communicating hazards of chemicals was later expanded considerably to later include provisions for "community right to know" and involved public disclosure of the chemicals used at each facility that might be expected to have some sort of public impact in the event of an upset condition or significant loss event. As the industry continued to experience high profile losses, regulatory rulemaking continued at a rapid pace along multiple lines simultaneously. Perhaps two of the most comprehensive performance standards to influence the evolution of Operational Excellence have been the US Occupational and Safety and Health Administration's (OSHA) PSM of Highly Hazardous Chemicals standard (29 CFR 1910.119) and the EPA's RMP rulemaking. Both were authorized as a result of provisions within the 1990 Clean Air Act. The legislation was intended to work comprehensively and assure both workplace and public safety. OSHA's PSM standard emphasizes the management of hazards associated with highly hazardous chemicals and establishes a comprehensive management program that integrates technologies, procedures, and management practices. These processes had to be managed with extreme care to avoid process-related loss incidents. Similarly, the EPA RMP regulations, published per the Clean Air Act, established rules and guidance for chemical accident prevention at facilities that use extremely hazardous substances. The standard requires disclosure of certain information from facilities subject to RMP, which in turn helps local fire, police, and emergency response personnel prepare for and respond to chemical emergencies. Making RMPs available to the public was also intended to foster communication and awareness to improve accident prevention and emergency response practices at the local level.

EPA's Risk Management Program Rule has a different set of exemptions than the OSHA PSM standard; OSHA exempts some processes that EPA does not exempt, and vice versa. The rules have a certain amount of overlap, yet, there remain differences in coverage and applicability. Accordingly, it is necessary to carefully examine each facility individually to determine whether a process is a covered process under RMP even if it qualifies for an OSHA PSM exemption.

What Is Operational Excellence (OE)?

1

Keywords

Continuous improvement; Cost & profitability; Developing countries; Efficiency; Governance; Government; Investors; Labor relations; Lawsuits; Leadership; Legal; License to operate; Management; Performance; Process; Public trust; Regulation; Reliability; Safety management systems (SMS); Shareholder; Social responsibility; Voluntary protection program (VPP); Workers compensation; World class performance; Zero incidents.

"Excellence" is defined as a talent or quality which is unusually good and so surpasses ordinary standards. It is also used as a descriptive for a standard of performance. Operational Excellence (OE) is the systematic management of health, environmental and safety in an integrated manner applied across all facets of the business to achieve superior outcomes across all the operations of the organization. Some organizations have described it as the ability of the organization to achieve and sustain leading performance in reliability and efficiency, while adhering to the highest standards for safety, health, and environmental stewardship in a cost-effective and profitable manner. Regardless of the many different ways it has been described, we can tell you that OE is a journey, the sort of journey that defines an organization and the capability it nurtures in pursuit of world class performance and competitive advantage as opposed to just delivering them to a specific destination. Organizations choose to apply OE through a comprehensive, structured, and well organized approach to building organizational capability as well as to leverage benefits resulting from doing things the right way. Most organizations employ OE systematically using a continuous improvement approach. At the heart of OE is *leadership* and the on-going efforts of establishing and maintaining standards in a consistent and reliable fashion. This focus invariably leads to greater enhanced operating

Applied Operational Excellence for the Oil, Gas, and Process Industries. http://dx.doi.org/10.1016/B978-0-12-802788-2.00001-4
Copyright © 2015 Elsevier Inc. All rights reserved.

performance and efficiencies necessary to assure sustainability. And of course, getting this part wrong can have quite the opposite effect as well; failing to manage effectively to the point where losses occur should suggest the possibility for certain losses of efficiency and introduction of additional hurdles that only serve to diminish the potential for maximizing profitability. Taking the journey toward achieving Operational Excellence typically begins with making an initial step-change improvement, followed by a commitment to a continuum of incremental enhancements. Installing and nurturing a culture of Operational Excellence results in significant and sustained competitive advantage though safe, reliable and highly effective operations.

> *Operational excellence is no longer just something to strive for—it has become a "must have."*
>
> *Dr D.P. Nolan*

Companies striving for outstanding safety and health records choose to apply OE not only for ensuring strict regulatory compliance, but also for the purpose of developing their own best practices as a means of enhancing and proactively managing their own performance in a methodical and disciplined fashion. Ultimately, achieving operational excellence is about empowering all workers—*management, supervisors, employees and even contractors*—to make the decisions and take the actions necessary to make safety and health practices truly work. OE involves a strong focus on planning, decision-making, and taking the precautions necessary for preventing incidents that can result in injury or property loss. It simply makes more sense to figure out what can go wrong and take the steps to avoid it before it happens and avoid losses before they occur. This sounds simple enough in concept. But, with the urgent deadlines in today's daily activities, where there is the continual push to make things cheaper and faster, and to do more with less, it is no surprise that the many competing priorities an organization must contend with must be properly reconciled in order for things to be completed the right way—safely in a manner where there are *no incidents* and *no one gets hurt.*

A properly executed operational excellence management system will touch every level of the organization from the boardroom to the shop floor. OE requires management to engage the entire line management chain of command; mid-level management, front line supervisors, foremen and team leaders alike. Collectively, the line management chain of command share responsibility and are individually accountable for being good stewards of shareholder capital and properly managing the workforce, implementing and maintaining the system, and ensuring operational discipline across the spectrum of management system functional controls—comprised of the various processes, programs, procedures, rules, and safe work practices.

This book will examine key attributes that are part of the operational excellence journey and will also provide insights for developing a genuinely effective and sustainable company culture that truly values safety, health, and the environment within organizations in the oil, gas, and related industries.

Why OE and What's It All About?

Modern industrial businesses exist for many reasons but one of the primary reasons is centered around this formula many will remember from college course Accounting 101:

$$R - C = P$$

This is the formula used for determining profit. It is fairly simple and straightforward. The formula tells us that an organization's profit (P) is derived by taking revenues (R) and subtracting costs (C). The process of organizing revenue and costs and assessing profit typically falls to the organization's accountants in the preparation of the company's income statement. Typically, revenue will be the first line on the statement. The accounting department will take out the cost of goods sold to arrive at a gross profit. Taking it step further and backing out fixed costs and they arrive at operating profit. Once irregular revenue and expenses are considered we'll reach the bottom line net profit. We are interested in highlighting this very basic accounting review so as to reflect for a moment on the treatment of irregular expenses. Loss incidents that result in harm to people, property, and business interruption directly affect an organization's bottom line. Putting this into context, it is clear that a business case can easily be made for embarking on an operational excellence journey.

Enlightened management teams intuitively understand that operational excellence is centered on being prudent in both their words and actions by actively managing safety, health, and environmental performance in a comprehensive and integrated fashion. Success is critical, not only for ensuring strict regulatory compliance, but ultimately to effectively establish and maintain performance standards while remaining steadfastly focused on activities that lead to reaching commercial goals, providing reasonable returns to shareholders, and recognizing certain competitive advantages in the market place. Companies that run their operations in a controlled and disciplined manner—*one that avoids harm to people and property loss*—naturally recognize the benefits of greater productivity, less downtime, and fewer business interruptions, leading in turn to a greater ability to run the enterprise profitably. Operational Excellence involves systematically addressing continuous improvements through process and program enhancements for the sake of increasing management effectiveness and efficiency. Interestingly, modern accounting systems do not typically budget for property loss incidents or causing harm and injury to people. Matters involving professional ethics, personal prudence, regulatory compliance, and legal liability would prevent these issues from being treated as just another cost of doing business; life is precious and it is short, societal norms and laws demand responsible behaviors resulting in safe, healthy, and environmentally-friendly operations.

Most successful businesses approach operational challenges directly and commit to operating safely and responsibly and a key aspect of this includes avoiding loss incidents. Failure to do this properly may result in harm to people and financial losses that directly affect the bottom line. Because the idea of working safely is something that personally impacts each individual worker (no one is interested in sacrificing his own personal safety) asking any member of the workforce would confirm that working safely is something unilaterally expected of individual workers, management teams, and indeed entire organizations.

Companies have come under continual pressure—first from regulators and sustainability-minded nongovernmental organizations as well as from like-minded investors, customers, and employees—to disclose ever-increasing amounts of operational information related to management effectiveness. What were once indices and rankings created by special-interest groups are now detailed operational reports that provide important insights into a company's future financial performance.

Increasingly, many of the largest companies around the world publicly report business sustainability information. They report on matters ranging from how much water their operations consume to issues relating to supply chain stability, greenhouse gas emissions, and the personal and process safety performance and reliability of assets. All of these nonfinancial data can be used as leading indicators and insights by the investment community into mid-to long-term financial performance.

One of the fundamental drivers of OE is to diligently lead and manage the business in a manner that is aligned with established internal and external standards, while focusing efforts accordingly to achieve top tier performance. Doing so requires a conscientious systematic effort involving a continuous improvement approach, as well as a commitment to doing things the right way and the consistent effort of every member of the workforce. This is necessary, for the sake of each individual's personal safety and well-being, as well as to collectively meet commercial business objectives and ultimately bottom line profitability.

Today's investor contemplating asset allocation into a capital-intensive business (a business that requires a large amount of capital to operate) will thoroughly research a company's asset reliability, environmental performance, and risk management and safe operations processes, not simply its price-to-earnings ratio and earnings per share. Business interruptions, incidents involving injury or property loss, and regulatory violations all threaten business productivity and impact revenue, earnings, and shareholder value.

A company's heavy emissions of greenhouse gases (compared with its peers), for instance, could be the sign of an organization with potential for pollution reduction. It could also signal an organization that has significant cost-reduction potential via a more concerted management focus on projects to reduce energy consumption in the long term. Unplanned downtime, whether because of safety issues, mechanical failures, or poor supply chain planning, may similarly signal that a company will incur costly facility disruptions and expense. A well-orchestrated plan to enhance reliability, reduce unplanned downtime holds the potential for lowering cost and boosting productivity.

Conversely, the absence of such a plan can detract from enterprise value. Consider for a moment, the impact on a company's market value from industrial incidents. A 2009 research study titled "How Does the Stock Market Respond to Chemical Disasters?" by Gunther Capelle-Blancardy and Marie-Aude Lagunaz, examined 64 explosions in chemical plants and refineries worldwide between 1990 and 2005 and found that, on average, petrochemical firms experience a drop in market value of

1.3% over the two days immediately following a disaster. The study calculated that each casualty corresponded to a loss of $164 million. A toxic release corresponded to a loss of $1 billion.

For companies already struggling to meet the demands to provide sustainability and other operational data, the financial market's increasing interest in operating information may seem like just another burdensome effort and expense. In reality, it is an enormous opportunity. Financial numbers are strong short-term performance indicators, but they do not always tell the full story. To the trained eye, nonfinancial operational metrics can help complete the narrative, especially when it comes to projecting mid-to long-term performance.

It's as Much About People, as the Bottom Line

No person in their right mind puts their shoes on in the morning and heads off to work consciously committed to cause harm to themselves or others, yet injuries to people and losses associated with property damage still occurs regularly across the oil/gas/petrochemical industry and indeed across industry in general. Every incident, whether large and small, represents a failure of the management system to adequately control the circumstances involved with the work processes. These sorts of losses often end up being viewed as one of the irregular expenses mentioned earlier and come right off the bottom line representing an immediate obstacle or impediment to achieving profitability. While it is generally recognized that money can be spent by company management after the fact, to repair, replace or refurbish damaged property, it should obviously be clearly understood that no amount of money can bring a person back to life once a life is lost to an on job fatality. People represent a company's greatest asset, as it takes people to conceive ideas and tie them to the actions necessary to make things happen. Management leads the efforts of the company's workforce and working together—*people* either achieve operational excellence or fall short of their business objectives. Without the efforts of the men and women that make up the company workforce, commercial success is not possible.

Leaders that run their businesses efficiently and effectively, do not injure their workforce and make every effort to pro-actively communicate, coordinate, and organize so that risks are properly quantified with hazards identified and mitigated so as to avoid injury, property loss, and business interruption.

Management teams have an ethical and moral responsibility to properly manage their businesses, as well as a fiduciary responsibility to their shareholders to be good stewards of capital, as they go about the day-to-day efforts associated with organizing and running their business. Being good stewards of financial capital is important, and most recognize the importance of investing it wisely and protecting plant and equipment so it remains reliable and fit for service, but equally important is investing in and protecting the human capital of an organization as well. Above all, OE is about running a sound business in a safe and environmentally responsible manner where each member of the workforce is able to go home at the end of each day to their families just as they arrived—*unharmed, injury free.*

The Stakes Have Never Been Higher

License to Operate

A business invariably requires a *license to operate* in both the figurative and very literal sense. There are numerous legal requirements for business licensing, all necessary to operate various aspects of a business, which have their origins codified by law. Referred to as business or operating licenses or permits, these instruments are issued by government agencies to allow individuals or companies to conduct business within the government's geographical jurisdiction. It is the authorization to start and operate a business issued by the local, state or Federal government(s). A single jurisdiction often requires multiple licenses that are issued by one or more government departments and agencies. Business licenses vary between countries, states, and local municipalities. There are often many licenses, registrations and certifications required to conduct a business in a single location.

Regulatory license is usually clearly defined in scope and received at a specific time, by a recognized government authority. In contrast to this, social license is often (and correctly) viewed as something informal with no specific basis in law. Nevertheless, social license shares certain attributes with the legal regulatory license. Notably, the social license is generally seen as conditional and, while not mandatory as the regulatory license is, is increasingly regarded as a practical necessity—whether or not specifically regulated by government authority. For the most part, regulatory license does not offset or simply negate the need for and value of social license. Beyond the license needed to operate required in the legal sense (i.e., the requirements for such licensing codified in law and regulations [e.g., construction permits. wastewater discharge permits, air permits, etc.]), there is also the broader social aspect of the concept that involves, reputation, the public perception of the company and its social license. This can be equally if not more important than the legal license(s). And because in certain situations the public is directly involved with government authorization of the legal license/permits (i.e., through the public hearing process), being able to accurately portray the company as a good corporate citizen that operates to the highest standards in a fashion that warrants public trust is generally seen as a positive influencing factor helping to favorably sway decision making. Public perception is also very important to companies that are publically traded on major stock exchanges, corporate image and financial results all play into the perception of how well run a company is and how likely it is to succeed over both the short and long term. Companies that elect to pursue operational excellence typically establish and maintain their reputations as good corporate citizens and are usually welcome additions to the communities in which they operate.

A company's license to operate is also important because it:

- establishes a certain legitimacy for its presence and actions from a local community perspective;
- enhances trust by clearly demonstrating to regulators and others that the company is genuinely striving for good performance;
- gives regulators assurance that an organization is acting responsibly;
- minimizes the potential for costly delays in regulatory approvals due to opposition;
- assures shareholders and investors that the company is managing social and other risks associated with its projects and activities;
- protects the company's reputation in times of crisis.

Although the terms and conditions of the social *license to operate* are often very informal and even somewhat intangible, the license to operate can never be self-awarded, it essentially requires the collective activities of an organization continually demonstrate responsible corporate citizenship sufficient enough to engender legitimacy and earn the public trust—as well as have the expressed or implied consent of those affected. Even in situations where definitions are not clearly determined in the law, businesses are held to reasonable standards and are judged daily in the court of public opinion. Because these are all issues tied to public opinion, it can be difficult for a business to directly determine the precise level of risk prevention or mitigation it should engage in to meet environmental or social risk—stakeholders and those affected need to be involved to give credence to the idea of legitimacy. Some of these issues are fairly straightforward, such as when there is a clear regulatory mandate that provides precise specification standards, in other instances where a government agency has crafted a performance standard, the organization must then apply logical and reasonable standards that the company can readily defend in the legal arena as well as in the court of public opinion.

When it comes to role of private business in this process, how is it that companies earn the social legitimacy and earn the respect and trust of the public in such a way that they are able to turn such legitimacy into profitability? This is done by responsibly delivering our society's essential goods, while efficiently and effectively managing safety, health and environmental obligations in its operations. Can operational excellence and the good management this involves be something that directly translates to a competitive advantage? Absolutely! Can the converse of this be true as well? Certainly, and to be reminded of this fact one need only look to companies that have failed to live up to public expectations in these areas and permanently faded from view.

The Legal Process–National Regulatory Regimes and Regulatory Frameworks

Developed Industrialized Countries

Legislative bodies have responded to the actions of individuals, courts, and influencing groups such as those loosely referred to by the acronym NGO (nongovernmental organizations). In response to societal norms, customs, beliefs and overall general expectations, they have started hearings/public inquiries, and/or worked behind the scenes to promulgate new administrative standards backed by new laws and regulations (or influenced expansion of the scope and interpretations of existing laws) where there has been a perceived need for improvement.

In the United States, the Occupational Safety and Health Act of 1970 was implemented because the US Congress found that personal injuries and illnesses arising out of work situations imposed a substantial burden on and a hindrance to interstate commerce in terms of lost production, wage loss, medical expenses, and disability compensation payments. The act authorized the Secretary of the US Department of Labor to set mandatory occupational safety and health standards application to businesses and created the Occupational Safety and Health Review Commission for carrying out adjudicatory functions under the Act. In the years since its establishment, the Department of Labor's Occupational Safety and Health Administration (OSHA) has been synonymous with progressive establishment and enforcement of safety and health standards which have had worldwide influence.

Similarly, in the UK, the Health and Safety at Work Act of 1974 (HSWA) is the primary piece of legislation covering occupational health and safety in Great Britain. The Health and Safety Executive, with local authorities (and other enforcing authorities) is responsible for enforcing the Act and a number of other Acts and Statutory Instruments relevant to the working environment. The Act which largely reflected the recommendations of the 1972 Robens Report, introduced a broad goal setting, non-prescriptive model, based on the view that *those that create risk are best placed to manage it*. In place of existing detailed and prescriptive industry regulations, it created a flexible system where regulations express goals, principles, and objectives and are supported by codes of practice and guidance. The Act gave people a broad legal right to be protected from work-related risks and hazards. In general, the law imposed a range of duties upon employers, the self-employed, and employees as well as others such as designers, manufacturers, or suppliers of articles and substances for use in the workplace. Many of these derive from EU directives. Besides laying down duties, the law also gave the Health and Safety Executive (HSE) and Local Authority inspectors wide ranging powers; to prosecute and to issue notices halting dangerous work or requiring.

EU OSHA

The European Union version of occupational safety and health legislation came along a bit later than either the US OSHA or the UK HSE. When it was adopted, the EU Occupational Safety and Health (OSH) Strategy provided a common framework for coordination and a common sense of direction. 27 Member States now have a national OSH strategy and have adapted to the national context and key priority areas. As risks to workers' health and safety are broadly similar across the EU, the Union recognized it had a clear role in helping Member States to address such risks more efficiently and in ensuring a level playing field throughout the EU. This role is explicitly recognized in the EU Treaty, which gives the Union shared competence to encourage cooperation between Member States and to adopt directives setting minimum requirements to improve the working environment in order to protect workers' health and safety.

> *No one should have to sacrifice their life for their livelihood, because a nation built on the dignity of work must provide safe working conditions for its people.*
>
> US Secretary of Labor Thomas E. Perez

There was generally felt to be a lack of an explicit legislative competence within the European Union Treaty addressing the field of safety and health in the workplace until around the mid-1980s. Until then occupational safety and health was seen as an annex to market harmonization and the economic policies of the European Economic Community. The 1987 Single European Act was seen as a major step forward in that it introduced a new legal provision on social policy to the Treaty targeting "improvements," specifically aimed at improving the health and safety of workers in the workplace. With the insertion of this provision into the Treaty, the importance of safe working conditions took on greater significance in the EU. Moreover, the new Social Chapter prompted the European Commission to promote social dialogue between employers and labor representatives at a broader European level.

With the Treaty of Amsterdam in 1997, the legislative competence in the fields of European social policies was further strengthened by the incorporation of the social agreement into the EC Treaty.

A directive is a legal act provided for in the EU Treaty. It is binding in its entirety and obliges Member States to transpose it into national law within the set deadline. A directive enters into force once it is published in the Official Journal of the EU. EU directives on safety and health at work have their legal foundation in Article 153 of the Treaty on the Functioning of the European Union (e.g., Article 137 TEC), which gives the EU the authority to adopt directives in this field. A wide variety of EU directives setting out minimum health and safety requirements for the protection of workers have since been adopted. Member States are free to adopt stricter rules for the protection of workers when transposing EU directives into national law, and so legislative requirements in the field of safety and health at work can vary across EU Member States.

REACH

European Union Legislation Requires Companies to Provide Detailed Information on Chemicals

In recent years companies have made considerable efforts to comply with the new European Union regulation commonly referred to as REACH (for **R**egistration, **E**valuation and **A**uthorization of **CH**emicals). REACH was adopted by the European Parliament and the Council in December 2006 and entered into force on June 1, 2007. REACH has several different implementation deadlines and it has also become legislation within the European Economic Area (EEA) countries (Liechtenstein, Norway and Iceland).

REACH is considered by some as one of the most complex and far-reaching Chemical regulations ever adopted by the European legislature and it has impact a wide swath of industry. According to REACH legislation, fuels, oils, waxes, lubes, and many gases are all chemicals subject to this legislation.

The legislation requires industry to mitigate risks from these chemicals by providing information on their properties and risks and registering that data with a central agency. In practice this effectively means that the responsibility for ensuring that chemicals are used safely is shifted from the regulatory authorities to the manufacturers and importers of the chemicals. Registration means that the producers and importers of substances and compounds are obliged to provide the European Chemicals Agency (ECHA) with a complex series of information regarding the characteristics of such substances and their uses.

A reading to determine the scope of REACH indicates it extends to stand alone substances, substances in mixtures (solution composed of two or more substances), and to some extent, substances contained in articles (the term articles refers to an object with a shape, surface or design that determines its function).

There are however certain exclusions from scope of the regulation which include: Non-isolated intermediates, radioactive substances, substances under customs supervision and waste material.

Examples of exemptions from registration and evaluation are:

- Substances manufactured/imported below 1 t/year (per manufacturer/importer).
- Annex V Substances including substances occurring in nature, if not chemically modified, e.g., Natural gas, crude oil.
- Polymers (but monomers and certain additives are subject to REACH requirements).
- Substances for Product and Process-Oriented Research and Development (PPORD). A notification is required; exemption period of 5 years (may be conditionally renewed).
- Substances used in products covered by specific end use legislation, such as medicinal products, foodstuffs.

The goal of REACH is to provide a greater level of protection for people and the environment by ensuring that proper information on handling, storage, and chemical make-up together with risk management measures, if appropriate, are readily available.

Leading companies recognize the importance of regulatory compliance, and check their own internal standards and adjust accordingly to be sure they remain well aligned with applicable legal obligations. The fact that many of the leading companies in the oil/gas and petrochemical industry systematically manage OE in a proactive responsible manner only serves to smooth the path to full compliance to the letter of the law. In many instances, through their systematic application of operational excellence, these industry leaders not only meet the spirit and intent of the minimums established by regulation but these companies often work to exceed these minimum expectations established by regulation.

The formulation of priorities and strategies in occupational safety and health is not undertaken by national authorities alone. In some countries, the setting of priorities and strategies is carried out in agreement with the social partners. In many other countries there is some form of consultation with the social partners. In some countries, autonomous regions are also involved in setting up priorities and strategies.

Developing Countries

With the globalization of commerce over the past several decades and the transformation of formerly agrarian economies into burgeoning new industrial based economies, (e.g., countries in SE Asia) there has been a strong focus on occupational safety and health standards in many of those countries in addition to those in the industrialized western countries such as the US and UK. As lesser developed countries welcomed manufacturers from the industrialized western nations, there was an expectation that the large conglomerates with responsible management teams and stellar reputations would also import and integrate their high standards for excellence—including the standards for exemplary safety health and environmental performance—as they expanded or relocated their production to foreign shores. As globalization escalates, more consumers and customers expect organizations to be ethical in every aspect of their business including the way they treat their employees. In recent years the media has exposed a number of organization's malpractices, leading to a significant impact on their brands and the loss of confidence in their business.

Over time, the regulatory standards in many of these (formerly) less-regulated countries have upped the ante and included their own occupational safety and health standards. NGOs such as the International Labor Organization (ILO) have worked right alongside them to promote a framework designed to provide coherent and systematic treatment of occupational safety and health issues and promote recognition of existing conventions on occupational safety and health. The aim of such organizations has been to establish and implement national policies on occupational safety and health through dialogue between government, workers, and employer organizations while promoting a national preventive safety and health culture. Operating organizations that were prepared and willing to bring the higher operating standards that exemplify operational excellence typically have fewer longer term challenges in dealing with a burgeoning regulatory regime in developing countries as they expand their business into those regions of the world. In contrast to this, companies that sought to close down heavily regulated operations in their home countries with a mature regulatory regime in hopes of achieving greater profitability by avoiding the obligations associated with safety and health regulatory requirements have or will soon find themselves disappointed as regulatory requirements expand, develop, and in a sense reach some sort of equilibrium and uniformity in requirements around the world.

Philippines

To date the ILO's Promotional Framework for Occupational Safety and Health Convention (No. 187) has been ratified by more than 30 member states. One such state is the government of the Philippines who now has their own version of OSHA and the Health and Safety Executive and it is referred to as the Occupational Safety and Health Center (OSHC). It was formed with Executive Order No. 307, on November 1987 by then President Corazon C. Aquino. The OSHC was established to be the national authority for research and training on matters pertaining to safety and health at work. Its mission is to provide expertise and an intervention mechanism to improve workplace conditions in the Philippines.

Thailand

The Thai Labor Protection Act B.E. 2541 (1998), prescribes OSH protection measures which are structured through the collaborative input of representatives on a national committee on occupational safety, health and working environment. It is comprised of representatives from each party; the government, employers, and employees. The committee works with the Minister to promulgate ministerial regulations and notifications in connection with implementation of the Act. The Minister is granted authority to issue ministerial regulations establishing workplace standards for employer implementation. The Act applies to workplaces with one or more employees.

On July 10, 2009, the Thai Cabinet approved the Occupational Safety, Health and Working Environment Bill (the "Bill"). The Bill sets out the government's long-term policy on health and safety in the workplace. The Bill updates the *Labor Protection Act* and adds new provisions to the eleven year-old Act. These changes aim to elevate the standards of occupational safety for workplaces throughout Thailand. Notable changes include new requirements including:

- *Employer Responsibilities*
  Employers must maintain safe, healthy and environmentally-friendly working conditions. They are also required to provide workers with training and information on avoiding occupational injuries.
- *Registration of Safety and Health Service Providers*
  Persons who provide examinations, verifications, risk assessments, training and consulting on occupational safety and working environment are now to be registered with the Department of Labor Protection and Welfare.
- *Safety Inspectors*
  These new officials are empowered to inspect workplaces, check machinery and equipment, collect evidence, and interview relevant persons to verify the working environment of a company. Safety inspectors can order employers to improve working conditions, or repair or stop the use of unsafe or dangerous machinery or equipment.

China

The organization formerly known as The Chinese Society for Science & Technology of Labor Protection (CSSTLP) changed its name to The China Occupational Safety and Health Administration (China OSHA or COSHA) in November 2003. Its multitask mission includes safe guarding the legal rights and interests of OSH practitioners by promoting industrial self-regulation, and protecting the safety and health of workers through compliance with laws, regulations, and social ethics.

With economic globalization and the many noted shifts in manufacturing centers over the past several decades, there has also been increased complexity in trade between countries. As this has coincided with the advent of an age of information technology that has allowed an unprecedented ease in international communication, governments around the world are now better equipped than ever before to see and understand what their counterparts in other government agencies do to regulate industry. Against this backdrop, it is easy to understand how the efforts of the United Nations (UN) to establish regulatory frameworks to be adopted by cooperating countries will continue to level the playing field throughout industry over time and serve as one means for establishing certain minimum standards and international uniformity—particularly in the area of safety and health. One recent example is the Globally Harmonized System of Classification and Labeling of Chemicals (GHS).

The GHS is an internationally agreed-upon system, created by the United Nations. It is designed to replace the various classification and labeling standards used in different countries by using consistent criteria for classification and labeling on a global level.

Before GHS was established by the United Nations, there were many different regulations on hazard classification in use in different countries. Although many were at least somewhat similar in content and approach, without a common framework, the regulations governing industry represented multiple standards and classifications and labels for the same hazard in different countries. Given the extent of international trade in chemicals, and the potential impact on neighboring countries when controls are not implemented, this created many unnecessary obstacles and trade complexities. Fortunately, it was determined that a worldwide approach was necessary to streamline and harmonize the rules.

The GHS was designed to replace all the diverse classification systems and present one universal standard which all countries should follow (however, note the GHS is not compulsory under any UN treaty). The system provides the framework and standards for participating countries to implement a hazard classification and communication system, an advantage for many of the less economically developed countries that may not have had the money to create one themselves. In the longer term, the GHS is expected to improve knowledge of the chronic health hazards of chemicals and encourage a move towards the elimination of hazardous chemicals, especially carcinogens, mutagens, and reproductive toxins, or their replacement with less hazardous ones.

GHS development began at the United Nations Rio Conference in 1992, when the ILO, the Organization for Economic Cooperation and Development (OECD), various governments and other stakeholders met at a United Nations conference. It supersedes the relevant European Union (which has implemented the United Nations' GHS into EU law as the Classification, Labelling and Packaging (CLP) Regulation) and United States standards.

In addition to the regulatory frameworks established on the international stage in both industrialized and developing countries, many readers will already be very familiar with the considerable domestic safety, health and environmental rulemaking that currently exists within the United States at the Federal level as well as in many instances will extends down to the state and local levels. Most states with individual state plans tend to follow the lead of the Federal OSHA Administration, and while a state has the latitude to exceed the Federal standards, it is typically understood that they cannot be any less stringent than the Federal requirements. California, with its CAL-OSHA is often held up as one of the most stringent state regulatory environments to do business in, but the reality of that notion can be more of a perception than a hard fact. Nevertheless, some will point to the widely publicized rules in the State of California that are popularly known by titles such as *Be A Manager—Go To Jail* or *Tell a Tale and Go To Jail.*

California's Corporate Criminal Liability Act: Be a Manager—Go to Jail

According to the Encyclopedia of Business, the chain of command, sometimes called the scaler chain, is the formal line of authority, communication, and responsibility within an organization. The chain of command is usually depicted on an organizational chart, which graphically illustrates the superior and subordinate relationships in the organizational structure. Classical organization theory suggests that the organizational chart allows one to visualize the lines of authority and communication within an organizational structure and ensures clear assignment of duties and responsibilities. The principle of unity of command is maintained, by utilizing the chain of command and its visible authority relationship. There is a school of thought that holds the idea of unity of command to mean that each subordinate reports to one and only one superior. While this may be accurate in most situations, there are organizations utilizing a matrix management approach that may result in individuals having more than one boss.

We will use the business term "Line Management" to broadly describe the administration of activities that contribute directly to an organization's output of products or services. In terms of corporate hierarchy, the line manager is an established authority within a vertical line (aka chain of command). With both authority and resources, line management is also responsible for establishing and adopting (under the leadership and with the support of senior management) an effective environmental, health and safety culture within the organization.

Senior leaders must be aware of the various regulatory requirements in the countries where they operate. Many countries have legislation which:

- Places specific legal duties on boards, organizations, and individuals in relation to the prevention of major accidents, and
- Incorporates sanctions such as corporate manslaughter when there have been serious management failures leading to a fatality or even other criminal charges for instances involving significant environmental damage.

Regulators around the world are increasingly focusing on the most senior level in an organization's hierarchy when trying to determine where the ultimate accountability for an accident should lie.

One example can be seen in the United States with the State of California's Corporate Criminal Liability Act (Penal Code Section 387). It provides that anyone "who is a manager with respect to a product, facility, equipment, process, place of employment, or business practice" is guilty of a misdemeanor or felony if they: (1) have actual knowledge of a serious concealed danger; and (2) fail to report that danger to the affected employees and the Department of Industrial Relations (DIR). Violation of the Act is punishable by up to three years in state prison and a $25,000 fine. With very real and serious consequences for failing to properly manage safety, health, and environmental efforts, there are often considerations for both monetary and criminal liability at stake. Those wishing to avoid both find it is far better to over-notify and risk a possible suit from a litigious employee than to remain silent and risk jail. This is likely the exact sort of paranoia the CCLA was intended to create. While safety is often considered its own reward, it is also something so critically important that it is not a matter left to chance. Yet, there are those in industry who are willing to play it fast and loose with the rules and willingly take on such risks either through their deliberate actions or through willful mismanagement or even through the act of omission. It is indeed unfortunate when accepting those risks that result (knowingly or unintentionally) in unnecessary injury to workers engaged in efforts on behalf of those signing their paychecks. Delivering subpar safety performance that clearly fails to meet the employer's mandate to provide a safe and healthy workplace in the State of California means that under CCLA, responsible managers could be subject to both civil and criminal penalties. The threat of jail time has proven to be a very powerful motivator for some senior managers and has influenced many operational decisions.

In contrast to those types of exceptional or otherwise clearly egregious situations, organizations that choose the Operational Excellence path continually improve their efforts and will typically avoid many of the problems and pitfalls that come with failing to meet the employer's obligations to provide a safe and healthy workplace. Getting this right allows management teams to take pride in their efforts and benefit from the operational efficiencies possible while avoiding the stigma of going to jail, paying large fines for poor management practices, or sullying the organization's reputation.

Toxic Spills and Toxic Torts

The organization's management systems are established to facilitate proper management of the business. When an organization fails to manage their business properly and such failures result in undesirable events such as fires, explosions, or release of hazardous materials in a manner contrary to established government law and regulations, the company may find itself defending itself not just from civil and/or criminal prosecution for such violations but also defending against civil proceedings resulting from claims arising from neighbors and others who may have been harmed as a result of the incident. In the US Federal rules of Civil Procedure, Rule 23 has been interpreted to permit certification of a "class" and when appropriate allowing for an action to be brought or maintained as a class action with respect to particular issues.

Trial lawyers have carved out a very lucrative niche for themselves and a number of firms have extracted damage settlements reaching millions and even billions of dollars. Toxic tort lawyers are quick to step in and represent themselves to potential clients as having substantial experience in responding to and helping their clients manage emergency situations where rapid response, strategic insight, and litigation know-how are critical. For example, law firms that have organized and moved in quickly following reports of an incident have learned that it helps to counsel their clients in the earliest phases following plant fires, explosions, chemical leaks, spills, releases, and groundwater contamination. To be effective and gain sufficient support for class action lawsuits they have inserted themselves into community relations, interacted with media, coordinated with emergency response teams, and even addressed themselves to issues related to insurance coverage.

Increasingly, toxic tort litigation problems involve not only civil litigation, but also interaction with state and federal regulatory agencies and criminal authorities. Many large law firms specializing in toxic tort litigation will have experience in all phases of the regulatory and criminal–litigation interface. They commonly negotiate on behalf of their clients with agencies and prosecutors. It is not uncommon for them to participate on behalf of clients at hearings and negotiations with government agencies such as the US Department of Justice, many State Attorneys Generals, the Consumer Product Safety Commission, the United States Environmental Protection Agency (and state environmental agencies), the US Department of Interior, the US Department of Energy, the United States Food and Drug Administration, the US Department of Agriculture, US Department of Interior, US Department of Energy, the Federal Trade Commission, the Occupational Safety and Health Administration, the Federal Aviation Administration, the National Highway Transportation Safety Board, the Federal Railroad Administration, and many other state and federal agencies.

Example Situations Resulting in Class Action Lawsuits

- A pipeline company defended itself in a multistate class action involving claims for property value diminution and personal injuries due to alleged polychlorinated biphenyl (PCB) spills.
- In connection with claims connected to groundwater contamination, a refinery in Indiana found it necessary to defend itself in a class action suit seeking damages for personal injuries, property evaluation, and medical monitoring following a flood that caused groundwater infiltration into homes. The defense eventually prevailed following a lengthy and expensive jury trial with a verdict that was preserved through an appeals process that ended in the US Supreme Court.
- A petroleum company defended itself in a case involving federal and state class actions arising out of groundwater contamination at a former refinery site. Plaintiffs sought injunctive relief, punitive damages, and compensation for alleged lost property value and impaired regional economic development.
- A manufacturing company had to defend itself in more than 70 suits filed in the aftermath of a tank car explosion in Alaska.
- A lead-paint manufacturer defended purported class actions potentially involving tens of thousands of plaintiffs seeking medical monitoring costs.
- Three large asbestos manufacturers defended themselves against a nationwide class of schools for the costs of removing asbestos from school buildings.
- A carbon black manufacturing company defended itself against multiple class action lawsuits in Alabama and Oklahoma brought by cities, businesses, and property owners alleging property damage from particulate emissions from carbon black facilities.

- A chemical manufacturer defended itself against thousands of formaldehyde lawsuits across the United States and Canada, including ten consecutive defense verdicts in the US and another in Canada following a trial that extended over seven years.
- A food and textile company defended itself in connection with a major toxic tort case involving alleged contamination of public drinking water with Cryptosporidium.
- A large oil company defended itself in connection with a class action suit brought by neighboring property owners for diminution in property value due to alleged air and groundwater releases into the community from its petroleum terminal.
- A large utility defended itself in connection with a suit brought by residents and owner of municipal water supply alleging that chemical discharges from a facility had contaminated the community's drinking water.
- A pharmaceutical company defended itself in multiple nuisance, trespass, and property damage claims arising from a 2008 flood that caused a wastewater pond to overflow into a creek, allegedly causing PCB-laden sediment to be deposited onto neighboring property.
- A pharmaceutical company defended itself against neighboring landowner's claims related to the company's ownership and operation of landfills. At issue was alleged damage resulting from groundwater contamination and underground debris.
- An oil company defended itself in phased trials following discharge of some 11 million gallons of oil into an Alaskan sound.

Organizations fully engaged in the Operational Excellence journey tend to be extremely well run operations and have established management systems with effective processes, programs, procedures, and internal standards to address issues that pose such risks as presented above. Through the application of risk management principles and sound operational discipline practices, they are often able to avoid many such pitfalls.

Large Scale Project Financing Dependent Upon Meeting Environmental, Health and Safety Guidelines

Getting Operational Excellence right and having an established track record for delivering world class safety performance results can be a real plus when it comes to demonstrating capabilities to potential lenders for the sake of positively influencing large scale project financing. To this end, proven operational excellence results can help secure outside project funding. Making such project financing contingent upon meeting these standards is one way of assuring operational excellence that in turn helps avoid losses and assure repayment. Astute lenders understand that it is extremely important to get safety right; after all, burning down or blowing up operating facilities can wreak havoc on the best plans or intentions to run a business profitably and without property losses or business interruptions. Lenders that provide financial capital want assurance that they will get a return *on* their investment as well as a return *of* their investment. Along that same line of thinking, it makes more sense than not, to protect the safety and health of the company's workforce. This holds true, if for no other reason than it takes people to make things happen and it is only through the efforts of people (workforce) that commercial success is even possible. Many companies recognize their workforce as their most important asset. These are some of the reasons organizations typically find a multitude of safety, health, and environmental stipulations to funding for projects—and this can be particularly demanding when these involve border-crossing international projects. One example is the comprehensive Environmental, Safety and Health Guidelines established by International Finance Corporation (IFC). These standards are commonly applied to large international scale projects financed by the World Bank.

When host country regulations differ from the levels and measures presented in the IFC Environmental Health and Safety (EHS) Guidelines, projects are typically required to achieve whichever is more stringent. If less stringent levels or measures than those provided in the EHS Guidelines are appropriate in view of specific project circumstances, a full and detailed justification must be provided for any proposed alternatives through the environmental and social risks and impacts identification and assessment process. This justification must demonstrate performance levels from alternate approaches will align with these standards and meet performance expectations of the host country and the lenders. Again, companies already moving forward on their own Operational Excellence journey have a much easier time of meeting and exceeding these operating challenges.

Social Safety Nets

The idea of social safety nets has evolved over time and these are intended to serve society in general by helping that part of the population that require some sort of assistance. Social safety nets such as Workers Compensation and Social Security disability benefits are two forms of compensation available to workers in the United States. These help to provide at least some form of the social safety net providing a modest income stream to persons who become disabled.

Prevention Efforts before *Recovery, Rehabilitation, or Restitution*

In a perfect world, everything would always work properly and no one would ever get hurt. But the real world we live in is still far from being a perfect place. Things don't always go as intended and when the undesired event occurs and someone gets hurt it is generally too late to do anything but make the best of a bad situation and make the effort to learn from what went wrong and take steps to make sure it does not recur again in the future. When property is damaged or people get hurt or disabled there is often a sense of urgency to ensure that people get the proper medical treatment followed by all necessary rehabilitation so that they can become productive citizens again as quickly as possible. In certain circumstances this can take varying lengths of time to achieve. When management teams reflect after the fact on what went wrong in an incident involving injury or property loss and begin to take the steps to avoid recurrence, at times it becomes obvious that avoiding the circumstances leading up to an incident might have been possible had only certain critical factors leading to the incident been addressed differently on the front end. And in retrospect, with the benefit of hindsight, many times it will also appear to be more cost-effective to spend the time, effort and money upfront in prevention efforts than to pay later for things such as recovery, rehabilitation and restitution. For this reason, progressive management teams will harness the value Operational Excellence brings to the organization by providing the necessary focus up front and approaching their work with a planning mindset and provide the proper focus on issues related to planning, engineering design (e.g., inherently safer designs), and important factors related to achieving operational discipline throughout an asset's life cycle. This involves assuring proper execution of operational and maintenance tasks, both large and small, and managing in such a way that affirmatively and pro-actively instills the discipline necessary to make sure things are always completed the right way so as to avoid the many undesirable events and assure reliable uninterrupted operations.

Rehabilitation before Compensation

Workers Compensation is typically mandated at the state level and is designed to provide compensation for workers who become injured on the job and the Social Security disability administered at the Federal level is another form of disability payment to help protect injured and disabled individuals. While they provide some means of interim financial support for those who are either temporarily or permanently disabled, they typically represent a significant reduction in income for the injured or disabled person over the course of his disability. For example, the Social Security Administration imposes limitations on benefits. In the United States a person cannot receive more than 80% of their average current earnings before they were disabled in terms of the combined amount of Workers Compensation and Social Security Disability benefit. While many of the developed industrialized countries often have such social safety nets in place, the type and extent of coverage tends to vary by region. In most circumstances the value of the compensation through these social safety nets does not measure up to the level of compensation an individual would otherwise earn through the normal course of his employment had he not become disabled through an industrial disabling injury.

The substantial impact workplace injuries and illnesses can have on income inequality was highlighted in a report by OSHA. They concluded that "Employers must do more to prevent injuries." Despite the decades-old legal requirement that employers provide workplaces free of serious hazards, every year, more than three million workers are seriously injured, and thousands more are killed on the job. The report states that these injuries can force working families out of the middle class and into poverty, and prevents families of lower-wage workers from attaining greater economic opportunity. Workers generally don't go to work in the morning with the intention of deliberately causing harm to themselves or their coworkers. Yet, at those times they find themselves involved with an incident resulting in injury at the workplace, it is fortunate that these social safety nets are in place to help provide some means of economic relief for the injured worker. Who bears the cost of worker injuries? OSHA estimates that upwards of 50% are out of pocket costs the injured worker must bear, 13% covered by private health insurance, 5% covered by State and Local Government, 11% by the Federal Government and 21% by Workers' Compensation insurance.

Workers' Compensation

Workers' compensation provides a just and fair means of compensating employees who have sustained injury, illness, or death during the course of employment. It is a kind of no-fault insurance. Injured workers may receive benefits without having to prove that their employer was negligent. The problem with fault and negligence as the basis of recovery for employee injuries is that the defenses available to employers were extremely difficult to overcome, leaving many injured workers uncompensated. Modern workers compensation regulation took fault and negligence out of the equation for compensation and made employers responsible for injured worker benefits without regard to fault. State law fixes the amount of compensation in a workers' compensation claim. The compensation is paid without consideration of fault or negligence on the part of either the employer or employee. In exchange for this right of compensation, the employee gives up the right to sue the employer.

The benefits an injured employee is entitled to receive are specified by law, so there is nothing voluntary about them. The insured employer's duties in the event of a loss include notifying the insurer promptly of all legal notices or demands as well as providing immediate medical services, cooperating with the insurer, and immediately notifying the insurer of a work-related injury. Typically, employers' liability insurance and workers' compensation insurance are found in the same policy but they are not the same coverage. Workers compensation benefits are paid to workers who suffer job-related injuries or diseases without consideration of fault or negligence. Workers' compensation is regulated at the state level. All states but Texas have compulsory workers' compensation laws. In these states, every employer must provide the benefits and amounts stipulated in the laws or face penalties for noncompliance.

Prior to the advent of workers' compensation, most employees injured on the job went uncompensated, creating an economic burden for themselves, their family, and society. Workers' compensation was introduced to provide a remedy for this situation. Workers' compensation laws provide that injured workers are compensated for their lost wages, the cost of covered medical expenses, death benefits, and the cost of rehabilitation. Unfortunately, the laws do not provide for retirement plan contributions.

Workers' compensation, which takes the form of a specified monetary benefit to the employee after a work-related injury, is paid without any consideration of fault or negligence on the part of either the employer or employee. Because workers compensation is structured by law to be an exclusive remedy, this means that the employee cannot sue the employer for work-related illnesses or injuries. It is also a certain remedy, for which the employee gives up the right to sue the employer for what could be a larger but uncertain benefit. Workers' compensation laws apply to most occupations. State workers compensation insurance does not cover workers in every category. What about the management teams of the large multi-national company with operations both on and offshore? In most cases, an executive of a company is considered to be a covered employee under workers' compensation law. Federal law covers all the other workers. Crews of ocean-going vessels are covered by the Jones Act, and railroad employees are covered by the Federal Employers' Liability Act. Harbor workers are covered by the Longshore and Harbor Workers Compensation Act.

Organizations embarked on the Operational Excellence journey often have lower employee incident rates, greater worker productivity, and enjoy better working relationships between management and the workforce. And because fewer disabling injuries and property loss incidents help the bottom line in terms of profitability, these companies often also enjoy lower insurance premiums (e.g., workers' compensation, property, business interruption) as a benefit of investing the time and resources necessary to operate in a safe and healthy experience modification factor is applied as a method of rewarding those employers with better than average loss experience and penalizing those with poorer experience. An insured with a good loss record pays a lower premium, and one with a poor loss record pays a higher premium. Premiums charged are based on the type of business (work or job classification) involved, the number of employees, and the total payroll. The rates vary greatly depending on the employment classification.

Labor Relations

Trade Unions

Organized labor, individuals freely associated for the purpose forming a union of common interests—often referred to as trade unions—collectively bargain with management and in so doing strengthen their efforts at influencing and improving working conditions, working hours, wages, job contracts, social security, as well as workplace health and safety. Cooperation between workers, their union representatives, and company management has been largely responsible for a great many such improvements over time and this was accompanied by progressively greater legislation and regulation and affecting workers' health and safety. Labor union promotion of safety and health issues throughout history is replete with very positive advances spanning the, socioeconomic, ethical, gender- and ethnicity-related, and even political boundaries. Through much of the twentieth century, and over the course of the economic boom in the United States following World War II resulting largely from a tremendous credit expansion, the unions have played a pivotal role influencing change in the industrial workplace. Wielding tremendous social and political clout, they are known for staunchly defending and supporting working conditions and being responsible for gradually strengthening the hourly workers position for the sake of achieving better pay and safer working conditions for their members.

Today, in the face of unprecedented rapid globalization, in the wave of free market capitalization sweeping the globe, those unions that were once able to secure lucrative contract concessions from companies established in the so-called industrialized countries in North America and Europe have found themselves a bit humbled as companies have shifted manufacturing operations to foreign shores in favor or lower cost labor.

As the labor market changes, the traditional membership base of trade unions in many countries has eroded. The increasing number of contract and outsourced workers has replaced a large number of permanent workers who were once members of trade unions, and this has undermined the bargaining power of many trade unions. In addition, economic changes resulting from so-called de-industrialization have also raised the number of workers in the informal sector. Workers in this category often don't have the same opportunities as organized labor pursues through collective bargaining.

Yet even against this backdrop, the influence of workforce organization and their trade unions can still be felt. The presence and strength of organized labor continues to grow and expand in developing nations much like they did in the US during the unprecedented growth seen over the last century. And it can reasonably be expected for this trend to continue—perhaps for extended periods of time (generations) as globalization and free market trade capitalizes on wide ranges in the cost of labor in various locales around the world as they continue the long march towards convergence and eventually reaching some sort of equilibrium. As this process occurs over time, labor relations and the way companies work with and respond to organized labor might also be expected to follow the same course as it followed during the industrialization of the United States, where unions achieved significant progress in improving working conditions (part of which included improvements in safety and health) for their members.

For example, in Vietnam, the Vietnam General Confederation of Labor (VGCL), an affiliate of the Vietnam Fatherland Front, now serves as an umbrella organization to which all unions in Vietnam must belong. Now with a membership of 3.7 million workers, accounting for 10% of the total Vietnamese labor force, the VGCL is comprised of 46,750 unions. Approximately 600,000 employees out of an eligible two million are unionized in the private sector, resulting in a unionization rate of 30%. Roughly 350,000 workers in foreign-invested firms are union members. In 2001, the VGCL claimed to represent 95% of public sector workers and 90% of workers in state-owned enterprises.

The VGCL is involved in national policy-making and is a powerful player at the provincial level. Internationally, it is affiliated with the World Federation of Trade Unions (WFTU) and has relations with 95 labor organizations in 70 countries. Yet, while in another Southeast Asia country—Indonesia—the government has ratified the ILO Freedom of Association and Protection of the Right to Organize Convention, 1948 (No. 87) and the Right to Organize and Collective Bargaining Convention, 1949 (No. 98) as well as enacting several key laws and regulations allowing workers' right to associate, to bargain collectively, and to get protection on employment terms and conditions—efforts to improve the situation of informal workers, contract workers, and outsourced workers in Indonesia continue to be plagued by difficulties. According to ILO DIALOGUE working papers, similar collective bargaining and social dialogue continues across countries of Southeast Asia as well as other regions of the world. Although with that said, progress is not without its challenges and it occurs at varying speeds according to locale. Some have predicted that over time though organized labor in the developing countries will gain some of the very same improvements in working conditions as workers that were won through collective bargaining in North American and Europe in the last century.

The following questions often define the strategy of trade unions in addressing occupational safety in the collective bargaining process:

- What activities should trade unions do directly in the interest of labor safety?
- What are the activities specific to the trade unions that will make a difference?

- What activities should reasonably be performed jointly with other represented workers' organizations?
- What are the tasks and areas where it is the obligation of the trade union to elaborate a standpoint reflecting the interests of workers, to represent and validate this standpoint through the involvement in the preparatory process?
- What should trade unions do in the case of illegitimate conditions where there is no means of prevention?
- How should trade unions treat employers who ignore workplace safety rules?
- Should trade unions engage the services of safety inspectors? If so, how? What relationship shall trade unions develop with the government safety inspection authorities?
- How should trade unions communicate violations of regulations to the public?
- How should trade unions exercise their duty to members in the area of occupational safety? How best can they work with their members and management to detect unsafe working conditions where there is no prevention and how should they intervene to prevent people from getting injured?

Trade unions often see their primary tasks in the area of occupational safety including efforts to:

- Make management, government officials, and other stakeholders accept the necessity of workplace safety.
- Emphasize that workplace safety is an integrated function within the all employment tasks/ activities and that adequate care and precautions are necessary of their members at all times to avoid injury or loss.
- Acknowledge other stakeholders (e.g., other organizations representing workers' are also present in the workplace) besides trade unions (e.g., public works councils, employees' safety councils, public education officials).
- Adapt, align, and pro-actively model safe behaviors in everyday trade union practice so as to be compatible with the organization's HSE approach.
- Expand and elaborate on the national occupational safety strategy with specific provisions for individual workplace application.
- Be actively involved in regulatory rulemaking as it pertains to influencing safety and health rules affecting union membership.
- Work diligently to detect illegal conditions with no prevention and intervene to eliminate such conditions:
- Refuse to accept unsafe and hazardous working conditions or those detrimental to the health of workers or the environment. Such situations should not be tolerated—even on a temporary basis.
- Establish safety minimums to ensure the safety of the workforce and other stakeholders.
- Clarify the responsibility of the state, employers, and employees as necessary.
- Consistently provide follow-up activities to influence effective control of worker protection. Selectively use publicity to raise awareness and exert public pressure as necessary and appropriate.

Operational Excellence and safety in particular are issues of critical importance to both management and the workforce. Regardless of whether the workforce is represented by a union and organized into one or more collective bargaining units or not, it should be universally understood that achieving OE and safety objectives is something of a shared mutual interest. Because of this it is one of the issues that can be used to motivate and influence performance that impacts other areas as well. Interestingly, as if to underscore the importance of this notion and highlight its importance, we can observe how it is incorporated into the fabric of numerous regulatory requirements as well as special recognition programs sponsored by government authorities to encourage operational excellence. One need look no further than provisions of the US Department of Labor's Occupational Safety and Health Administration's (OSHA) rulemaking. To examine this issue from a couple of different vantage points, we'll look at two points involving the US Department of Labor's OSHA. OSHA has established the importance of employee participation in matters involving safety in the workplace in provisions of the Process Safety Management (PSM) standard which are codified in 29 Code of Federal Regulations at 1910.119(c) *Employee Participation* and in the special recognition program sponsored by OSHA referred to as the Voluntary Protection Program (VPP).

The PSM standard's paragraph (c) requiring employee participation includes provisions for employers covered by the standard to develop a plan detailing provisions for how these employers will consult with employees and union representatives on provisions of the standard and explain how the program standards are being applied by the workforce. The standard clearly contemplates the need for including considerations that have the employer consult with employees and their union representatives (as applicable) on all aspects of the standard as well as providing the means for access and clearly communicating information necessary for workers to understand employer provisions developed under the standard. It makes sense when you think about it—there is a logical need to clearly communicate and coordinate with members of the workforce so the employer ensures they clearly understand their roles and responsibilities so the employer's process safety management program safely achieves its performance objectives.

Similarly, the VPP program was established by OSHA to promote effective worksite-based safety and health performance. The provisions OSHA uses to evaluate prospective program participants has provisions that focus first and foremost on issues related to management leadership, authority and line accountability but then in turn they also address issues related to employee involvement. In the VPP, management, labor, and OSHA establish cooperative relationships at workplaces that have implemented a comprehensive safety and health management system. Approval to admit entry into VPP is OSHA's official recognition of the outstanding efforts of employers and employees who have achieved exemplary occupational safety and health. While the VPP evaluation of an employer's safety and health management system are focused on assuring all elements of the site's safety and health management programs met the high quality expected of VPP participants (e.g., per Section IX "Areas of Excellence" of the evaluation report), it's also insightful to highlight the critical need for an employer's VPP application to reflect the support of its employees. The VPP application instructions state in Section II that an application must also concur with the application or OSHA will not accept the application.

The VPP has established performance-based criteria for evaluating a formal safety and health management system, invites sites to apply, and then assesses applicants against these criteria. OSHA's verification includes an application review and a rigorous onsite evaluation by a team of OSHA safety and health experts.

Depending on the outcome of the rigorous evaluation process, OSHA may qualify sites to be part of one of three programs:

- **Star**: Recognition for employers and employees who demonstrate exemplary achievement in the prevention and control of occupational safety and health hazards the development, implementation and continuous improvement of their safety and health management system.
- **Merit**: Recognition for employers and employees who have developed and implemented good safety and health management systems but who must take additional steps to reach Star quality.
- **Demonstration**: Recognition for employers and employees who operate effective safety and health management systems that differ from current VPP requirements. This program enables OSHA to test the efficacy of different approaches.

In addressing the importance of the individual worker, whether represented or non-represented by a trade union, employers should intuitively understand that success in Operational Excellence hinges upon buy-in, cooperation and diligence of every member of the workforce. Regardless of the manner in which the worker is engaged by the organization and to the extent that these individuals are direct hire employees, employees of contract service partners, or transient short-term supplemental manpower, it's clear that management must earn and maintain the trust, respect, and cooperation of every member of the workforce to fully achieve its safety performance goals. With regards to working conditions commonly negotiated between management and labor through the collective bargaining process (working conditions, working hours, wages, job contracts, social security, as well as workplace health and safety), many issues related to health and safety are often easier to agree on simply given the fact that members of the workforce can agree they don't wish to suffer injury simply for the sake of earning a living and similarly, employers understand it's a good idea to avoid seeing their employees get hurt or suffer the losses that come with injury costs, property loss, and business interruption. In short, the safety component is foundational to Operational Excellence. It makes good business sense and as such is clearly a point of common ground in collective bargaining and issues requiring workforce uniformity and unity.

Responding to or Simply Staying Ahead of Shareholder Activism

Pressures external to the organization such as those associated with shareholder activism and socially responsible investing present common examples of why an organization needs a sharp focus on achieving operational excellence—those that get it right are better positioned to uphold their reputations within the industry and defend themselves against situations that often wind up being public and investor relations nightmares. The Sisters of St. Francis are an unusual example of the shareholder activism that has been making waves through corporate America since the 1980s. More traditional shareholder activists have been public pension funds, who routinely flex their financial muscle on issues from investment returns to workplace safety and environmental performance. Rabble-rousing hedge fund managers and mutual fund managers have been known to charge in an attempt to shame companies into replacing their C.E.O.'s, shake up their boards—anything to bolster the value of their investments. And it is not unusual for such activist groups to set their sights on those companies who fail to meet their social and financial obligations—particularly vulnerable are those management teams who deliver subpar health, safety, and environmental performance leading in turn to subpar financial returns to shareholders.

The nuns who represent the Sisters of St. Francis have taken a very direct approach to assuring the performance of their retirement funds and at the same time they use their actions to enhance the greater good. Using the investments in their retirement funds as the stepping stone to influence change they have effectively become Wall Street's moral minority.

Case Study in Shareholder Activism—the Sisters of St. Francis

The Sisters of St. Francis formed a community to take up the cause of assuring corporate responsibility. Troubled by what they saw in businesses where they had invested their retirement funds, they formed a committee to combat these undesirable corporate developments. Working in coordination with groups like the Philadelphia Area Coalition for Responsible Investment, they mounted an offensive against some of the largest corporations in America. They have effectively boycotted Big Oil and even "encouraged" Big Tobacco to change its ways.

Eventually, they developed a strategy combining moral philosophy and public shaming. Their mode of operation is really nothing remarkable except for the fact they use channels available to all common shareholders. There are two main channels through which shareholders engage the corporation—this is through shareholder activism and socially responsible investing. They would simply take an ownership position in their target company, purchasing the minimum number of shares that would allow them to submit resolutions at that company's annual shareholder meeting. Interestingly, most companies decide they would rather avoid confrontation and simply let the nuns in the door than publically confront religious dissenters.

A spokesman for a firm specializing in shareholder proxy votes once observed that an organization is never going to get any sympathy for cutting off a nun at the annual meeting. And with their moral authority, the Sisters of St. Francis were thought to have the capacity to garner wide public attention to issues.

On March 24, 1989, a marine vessel owned by the former Exxon Shipping Company, bound for Long Beach, California, ran aground on the Bligh Reef offshore Alaska, resulting in the second largest oil spill in United States history. Following this high-profile petroleum industry environmental disaster in the early 1990s, Amoco Corporation was the target of a series of environmental shareholder proposals that were asking Amoco to sign the Valdez principles (a set of principles for environmentally responsible corporate conduct) developed by the Coalition for Environmentally Responsible Economies (CERES). While Amoco was not the company responsible for the disaster offshore Alaska, it was one of several industry players to be targeted with proposals regarding the Valdez principles (other companies included for example Exxon, Mobil, Dow Chemical), the way Amoco responded to the proposals and how the relationship between CERES and Amoco developed over time is a very interesting case study in the changing corporate approach to shareholder activism.

Initially the reaction by Amoco was similar to the hostile reactions of companies in the early days of shareholder activism. The Valdez principles were considered an unwelcomed, undesirable, and unwarranted intrusion into corporate affairs by an external group that has developed principles outside of the sphere of corporate influence and were even seen as a legal threat. Despite this resistance, the proposal achieved 8.6% approval at the annual meeting and thus triggering eligibility for reconsideration at subsequent annual meetings.

Interestingly, despite public opposition to the principles, Amoco began to shift its own environmental policies from compliance to a proactive approach, which was more in line with the Valdez principles than Amoco would admit. Rather, Amoco justified the internal changes with reference to efforts that had been under way for some time and based on benchmarking studies of competitor policies. In addition to these internal policy adjustments, Amoco agreed to meet with CERES and Friend of the Earth, who were sponsoring the proposals for the upcoming annual meeting, to begin dialogue about the aims of the proposals. As a result of the meeting and agreement by Amoco to continue meeting with CERES and offer accounting of their environmental activities, the proposals were withdrawn. Thus, "in dealing directly with this external source of the pressure, Amoco had opted for strategies that involved either avoidance or neutralization of CERES's efforts." This strategy evolved further and Amoco joined with IBM and subsequently additional companies (e.g., AT&T, Dow) to develop an alternative set of principles for environmental disclosure with the help of CERES. Amoco recognized that increased environmental reporting was a necessity in order to maintain a business advantage in the marketplace.

BP Shareholder Losses

The Deepwater Horizon was a deepwater, offshore oil drilling rig owned by Transocean (RIG) and contracted to the operator BP Plc. (BP). On April 20, 2010, while drilling at the offshore BP Macondo Prospect, there was a blowout on the rig

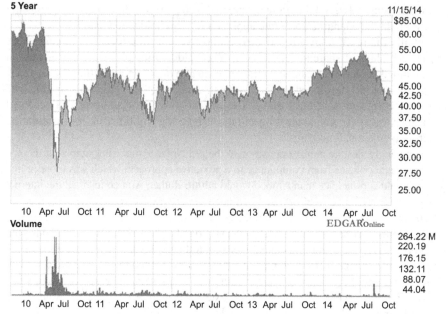

Figure 1.1 BP Stock price over 5 years.

followed by a release of gas and in turn an explosion that killed 11 crew members. On April 22, 2010, Deepwater Horizon sank while the well was still active resulting in the largest offshore oil spill in US history.

After the spill, the US Environment Protection Agency (the EPA) barred BP from bidding for new work in the Gulf of Mexico and supplying fuel to the military. The ban was lifted in March 2014.

The fire not only destroyed BP's physical property and caused casualties, but also destroyed shareholder wealth. From April 19, 2010, to June 25, 2010, BP's share price dropped 55%—from $59 a share to $27 a share (See Figure 1.1), and did not immediately recover up to the pre-incident price level and certainly did not compare favorably with its industry peers in the years that followed.

However, share prices did subsequently recover, although they never returned to pre-crisis levels and of particular note, they have simply not maintained parity with BP's industry peers over the four years following the incident. Between August 2010 and August 2014, shares averaged $44 a share. This average is 27% below the peak that shares reached just before the incident. At the same time, BP's industry peers realized between 76% and 87% gains in their share prices and reflected accordingly by respective increases in their market capitalizations.

The point here is that shareholders who entrusted their capital to the care of the BP management team were rewarded with subpar performance when compared to the performance of other organizations in this segment of industry over that same period of time. The verdict of the capital markets was clearly one which showed less confidence in the BP management team than other organizations competing within the industry.

These repercussions may seem of little consequence when compared to the losses suffered by the families of those who perished in the tragic Horizon incident, but the lasting financial impacts the BP organization felt in the aftermath of the disaster continues to put them at a disadvantage compared to their peers.

Why Corporate Governance Matters to Companies

Organizations are now routinely ranked by the investment community for performance in areas relating to corporate governance. Large institutional investors as well as the smaller individual so-called "retail" investor now have easy access to information on publically traded companies governance rankings/scoring metrics to help them guide their investing decisions so as to make informed decisions before they invest their capital. The Institutional Shareholder Services (ISS) Governance QuickScore is one such source that measures certain governance metrics with reports readily available to most through financial information resource providers such as "Yahoo Finance" (typically located on each company's Profile page, above the "Key Executives"). A quick glance at the ISS Governance QuickScore helps investors identify and monitor potential governance risk and to identify possible investor concerns based on signals of governance risk. The ISS positions itself as the global leader in corporate governance, with services that are aimed at enabling the financial community to manage governance risk for the benefit of shareholders.

Focused primarily at a top down macro level, such insights provide information into structural organization issues. These factors also play an important part in execution and the ability to achieve operational excellence. With a multitude of factors the ISS uses to evaluate corporations on their governance effectiveness they seem to have a reporting bias that at first glance may often seem more slanted towards the area of finance, but it would be unwise to dismiss the significance of the evaluation simply because it does not mention OE, social responsibility, or sustainability. There are of course numerous factors they evaluate that can quickly and easily be recognized as having a certain level of overlap—both directly and indirectly related or linked in some fashion (resources, controls, etc.) to operational excellence management system performance results. For example, factors listed in the QuickScore 3.0 Appendix (factors by region October 23, 2014) include the following:

- Audit and Risk Oversight
  - (5) Has a regulator initiated enforcement action against the company in the past two years?
  - (200) Has a regulator initiated enforcement action against a director or officer of the company in the past two years?
  - (201) Is the company, a director or officer of the company currently under investigation by a regulatory body?
  - (8) Has the company disclosed any material weaknesses in its internal controls in the past two years?
  - (125) Does the company disclose a performance measure for other long-term plans (for executives)?
  - (326) Did the company disclose the metrics used to evaluate performance-based compensation in the most recent Filings?

As part of the overall emphasis being placed on corporate governance at many levels in the organization, large international energy companies have increasingly turned to implementing comprehensive OE programs as the means for achieving what is portrayed as world class performance. And the way they incorporate it into their corporate governance efforts has an important impact on their ability to achieve and sustain world class OE performance and enjoy the benefits of an efficient and effectively run organization. These companies view OE as an absolute necessity for the long term added value it brings to shareholders and have integrated it accordingly into their operating models. In fact, one remarkable trend in corporate governance that has emerged over the past 25 years has been the rapidly expanding support for corporate and social responsibility (CSR) leadership as well as sustainability reporting. Interestingly, in many instances OE has been one of the primary vehicles to drive this effort so to a large extent these are all part of a same tight weave that works together to support the organization.

On the Growing Commitment by Corporations to Sustainability Reporting

It is our view that successful companies are those that see business objectives and sustainability objectives as interlinked.

ExxonMobil's 2009 Corporate Citizenship Report

The expansion of company CSR work and reporting is global and growing. In 2008, the consulting group KPMG reported that 79% of the largest 250 global companies produce CSR reports. Overall support for this trend is growing across all industry sectors around the globe. An Accenture study in 2010 shows that companies involved in the automotive, banking, mining, energy, and utilities industries see sustainability issues as very important compared to companies in the communications and IT fields. The business case for CSR is multifaceted. Because of the strong link between CSR and OE, it is not surprising that many of the talking points in support of both are very similar: Companies are increasingly reporting on their sustainability efforts. This suggests that if a company does NOT do the CSR reporting it becomes a vulnerability that give their competitors something of an advantage over them simply by the lack of transparency in this area, something being demanded more and more by the public.

There are several factors that would support visibly linking an organization's OE efforts with CSR, some of these include:

- Decrease Litigation and Risks Regulatory
  Global changes continue to stimulate new legislation and additional buildup of regulations. The advent of the internet has resulted in a broader awareness of expectations put on businesses in some locales and not in others and this prompts many governments to play catch up and develop their own rules so as to level the playing field for all. Over time this may ultimately result in convergence and a level playing field with respect to legal and regulatory requirements around the world. But for the interim, companies with worldwide operations

committed to CSR are often better equipped to face such challenges. For example, a company that has already set environmental and safety goals and objectives will be ahead of the crowd when governments late to the game of administrative rulemaking write new regulations that the company is meeting in another locale.

- Build Public Trust Avoid and Reputational Risk
  Trust is a crucial factor for long-term business prosperity considering many consumers make purchasing decisions based on their perceptions of company trustworthiness. A commitment to CSR reporting can help prove to stakeholders and consumers that a company is accountable and trustworthy. For example, the (Dell, 2009) Corporate Responsibility Report highlights this by stating "we must continue to build trust with customers and stakeholders by demonstrating our positive impact on society and the planet and developing meaningful measures for reporting our progress."
- Opportunities to Reduce Costs and Enhance Revenues
  Many companies have already found multiple ways to cut operational costs by utilizing their OE/sustainability programs. This is done by achieving operational discipline across the organization's operations, reducing waste and by avoiding operational incidents resulting in injury or property loss by improving workforce safety programs, processes and procedures. Leadership in CSR can also improve a company's efficiency, productivity, competitive edge, long-term survival and the ability to attract labor, investors and consumers. CSR reporting helps companies better integrate and gain strategic value from existing sustainability efforts and identify gaps and opportunities to enhance their revenue in operations.
- Advantages in Recruiting Labor
  A company that understands CSR issues and discloses its safety practices, non-discrimination policy, and worker benefits will be seen as a good employer and be better able to compete for top talent in its industry. Recalling a circumstance during the recent oil boom, we remember when a contractor crew of pipeline builders abruptly left the employ of a large contractor to one of the large integrated oil companies engaged in building a pipeline in Colorado. The reason they left? So they could earn an extra $0.50 an hour and enjoy a shorter commute to the new construction site of a different company. Interestingly, within a week's time the entire crew returned to the site of the project they left earlier and asked for their jobs back. When asked as to why the change of heart, one of the workers told the foreman that even though they would earn less and drive further to work (this also meant more personal fuel costs for their daily commutes) for this leading integrated oil company, the work crew unanimously agreed that they could see a contrast in each respective employer's approach to safety and operational excellence that the owner had crafted with the contractor and they frankly preferred to work at a job site where they could see that the company visibly and actively cared about worker safety and organized the work so as to assure they completed the entire project safely—they were simply more comfortable working there knowing the company's standards and the resulting work environment reasonably assured them that they would not get injured or killed on the job. The pipeline construction proceeded as planned and the crew remained on through the end of the job. This was a significant win because at the time there was a great deal of competition for trade and craft workers, so in this instance the company's OE management system proved its value in more ways than one.
- Advantages in Being Responsive to Investors
  OE management systems enhance efficiency and effectiveness of the company's efforts across the board. More effective safety management performance helps increase profitability through the avoidance of incidents causing injury, property loss and business interruption. It also adds to reliability over the long term. Increased savings and profit derived from corporate sustainability will obviously result in better shareholder returns. Moreover, for the long-term investor, a company with a forward looking view on managing sustainability issues is very encouraging.

Strong OE leadership and corporate governance is vital, because it shapes and molds an organization's corporate culture, and it is this culture which defines and influences employee behavior with respect to safety. Tasks involving safety may be delegated, but responsibility and accountability will always remain with the organization's senior leaders, so it is essential that they promote an environment which encourages safe behavior.

Using OE as an overall umbrella, senior leaders work to effectively integrate all the necessary components into their OE objectives, plans, processes, programs, standards, practices, and behaviors. This in turn gets woven into the fabric of their daily operational routines in order to protect people, the environment, and gain an economic and competitive advantage in their respective industries. These functions are all structured to work together so as to assure operational discipline and this coupled with other formal internal controls (i.e., financial accounting related) all serve to assure that the company effectively maintains its fiduciary responsibility to the company owners. These efforts can be very complex and challenging, but doing things the right way, results in operations that are prudently and responsibly managed in accordance with applicable standards.

> *Good things happen when we integrate sustainability into our products, services and solutions. We improve our competitiveness and create and capture customer value. We save money, reduce our environmental impact and improve our employee satisfaction.*
> *Chairman and CEO Caterpillar—Jim Owens*

It's easy to see how this can be a win–win situation for everyone involved. It should also be obvious that getting it right also serves to fulfill the organization's mission, vision, and values. This in turn helps to protect the workforce, other stakeholders and ultimately delivers the performance expected of company owners—the shareholders.

> *Creating a culture where all employees expect the unexpected and strive for error-free work is absolutely essential for success in process safety. This kind of culture is possible only through demonstrated leadership at all levels of the organization.*
> *Bob Hansen, CEO Dow Corning*

Good corporate governance of OE is not just about avoiding potential negative effects. There are a number of positive commercial reasons why an effective OE management system makes good business sense as well.

Some of the benefits of well managed assets and processes include:

- Improved efficiency and flexibility;
- Plants and equipment which have longer life spans;
- Less downtime, and higher plant availability;
- Maintenance budgets that are easier to forecast;

- Enhanced employee, stakeholder and government relationships and
- Ease of access to capital and insurance often at preferable rates.

These factors allow operations to run more smoothly and help to create a better, more productive business, with a less stressful working environment for managers and employees alike. Other on-going related benefits include:

- *Improving Stakeholder Relations.* Better corporate governance leads to transparency and better disclosure, thus providing the opportunity to establish relationships with all stakeholders in fair and more productive terms. In contrast, poor governance under mines the integrity of publicly traded securities and discourages the use of public markets as a means to channel investment. Even in less than extreme cases poor governance can serve to diminish the organization's opportunities for longer term growth and profitability.
- *Reducing Investment Risk and Adding Value.* Improving corporate governance presents opportunities to manage risks and add value to stakeholders. In bad corporate governance environments, poor adherence to standards and weak enforcement are barriers to investment and poor operational performance stymies returns. Improving the company's corporate governance allows them and commercial partners to invest in higher risk/high return environments. It can also increase the market valuation of companies and can attract further investment, which in turn would increase exit opportunities on favorable terms.
- *Avoiding Reputational Risk.* If companies do not work to improve the corporate governance, then they assume investment performance risk, as well as reputational risk. This risk is particularly serious when minority shareholders and other stakeholders stand to lose from governance abuses. It also influences the manner in which the company is more widely perceived in the public. For example, spending millions of dollars on an advertising campaign may be money and efforts wasted if the company has to overcome the sigma of a large scale operational catastrophe. Similarly, consumer alienation is possible when public perception of the company hinges on the bad press associated with operational failures in another area.

What OE Is and Is Not

Businesses, particularly those that involve the potential for major hazards, such as those in the oil, gas, and petrochemical industry, are often characterized as having the potential for large catastrophic accidents involving large scale loss of life, harm to health, and extensive environmental damage. There is no doubt that the manufacture of chemicals and petrochemicals, oil and gas exploration and downstream production, energy and power generation involves complex processes with intrinsic hazards that require diligent management. The measures needed to properly mitigate these hazards in a controlled way are equally complex and not always readily understood by the public, but they certainly should be by those responsible for bringing such products to market.

Safe operation and sustainable success of a business are inseparable. Getting OE wrong and failing to properly manage process safety detracts from sustainable positive operational results, and can cause delivery of sub-optimal financial rewards over the long term. And as we have already seen, the consequences of failing to properly manage major hazards can be extremely costly.

In today's business world there is far more to be gained through the pursuit of operational excellence than not. We have recounted several of the more high profile examples to highlight the need and benefit of effectively applied operational excellence. And while these highlighted only a few of the major consequences of getting it wrong, it's really only one consideration among many for deciding on the path to OE. While the outsized consequences of getting EHS performance wrong should be more than compelling enough to support the decision for OE, it's also important to highlight the positive side and emphasize the more important points relating to the benefits that come with getting OE right. To keep things in the proper context, it's important to keep in mind that getting it right takes much more than just putting words on paper with the ceremonial management pledge to signify support. It's not enough to just assume that the organization is achieving world class results because there are some cleverly crafted and openly displayed slogans or themes that proclaim the company values safety, health, and the environment, *the actions must match the words*. Commitment without follow through is hollow and often an exercise in futility. But where does one start? Getting OE right starts from the boardroom and involves leading from the top. Senior management establishes the vision and sets the tone that shapes the culture for the whole organization. The tone set at the top of the organization has a dramatic impact on the decisions and actions that follow and as the company's leaders lead by example, it sets the direction for the entire company. It is up to every member of the line management chain of command to extend accountability in their respective areas of responsibility for the safety management performance that yields the operational discipline necessary to assure success. OE is a journey of deliberate actions supported by continuous improvement requiring the diligent efforts of the entire workforce. It requires close coordination and communication and above all unity in purpose. The decisions made by subordinates in the line management chain of command will have a direct bearing on Operational Excellence and the consistent delivery of positive safety outcomes. Many companies, particularly those in high-hazard industry sectors have made important strides in establishing the necessary corporate culture and leadership to minimize the frequency and severity of safety and environmental incidents. So in other words, this is something that CAN be achieved; *if they can do it so can you*!

There is a saying that every journey begins with that first step. A conscious decision has to be made to begin that journey and the company's line management chain of command must be fully engaged to provide the leadership necessary to bring OE to life. As with anything requiring people to work together to achieve a common goal, the reality of achieving an effective operational excellence system is that it requires the buy-in and cooperative action of every member of the workforce. Operational Excellence serves no useful purpose if it is just dead paper on the shelf. Operational excellence is more than just a periodic audit of facilities and operations to measure a level of compliance. OE is largely about performance management and it is also a mindset. It helps keep everyone focused on doing the right things at the right time for the right reasons. The written commitments must match the actions and to do this the line organization must bring it to life and implement it effectively if it is to form the basis of the entire organization's decisions and actions and truly prove to be beneficial.

It is an overarching management system that integrates and incorporates elements such as leadership and accountability, risk assessment and management, communications, competency and training, asset integrity, safe operations, contractors suppliers and others, emergency preparedness, incident reporting and analysis, community awareness and outreach, and continuous improvement.

How Do Major Oil/Gas/Petrochemical Companies Describe the Basis for Their Operational Excellence Management Systems?

Safety, health, and environmental management are at the heart of Operational Excellence and while these three functions have quite a bit in common with each other, they also have much in common with quality management. Many of the key players in the oil, gas and petrochemical industry have embarked on their own Operational Excellence journey. The structure and approaches they each take will vary somewhat, but they share some common areas of emphasis and program features that they believe will help them deliver world class performance and achieve some degree of competitive advantage in their respective markets. Although many of these companies treat their Operational Excellence Management System (OEMS) materials as proprietary and in some instances restricted or confidential documents, through our industry interactions we have had a chance to review enough of them to recognize similarities and indeed some very strong parallels between them. The following provides summary overviews describing programs and approaches embraced by some of the key players in the industry.

We have examined OEMS systems from several major integrated organizations, which we will refer to as Companies A, B, C and D. This section examines how each of these have determined their approach and basis for their individual OEMSs.

Company A

- Company A describes their approach to operations integrity management in this manner: with this:

> *(Company A) is committed to conducting business in a manner that is compatible with the environmental and economic needs of the communities in which we operate, and that protects the safety, security, and health of our employees, those involved with our operations, our customers, and the public. These commitments are documented in our Safety, Security, Health, Environmental, and Product Safety policies. These policies are put into practice through a disciplined management framework called the Operations Integrity Management System (OIMS). (Company A) OIMS Framework establishes common worldwide expectations for addressing risks inherent in our business. The term Operations Integrity (OI) is used by (Company A) to address all aspects of its business that can impact personnel and process safety, security, health, and environmental performance.*

Company B

• Company B describes their operational excellence management system with:

> As a business and as a member of the world community, (Company B) is
> committed to creating a superior value for our investors, customers, partners, host
> governments, local communities and our workforce. To succeed, we must deliver
> world class performance exceeding the capabilities of our strongest competitors.
> Operational excellence (OE) is a critical driver for business success and a key
> part of our enterprise execution strategy. Operational excellence is defined as "the
> systematic management of process safety, personal safety and health, environment,
> reliability and efficiency to achieve world class performance."

To achieve and sustain world class performance, we must develop strong capability in operational excellence throughout (Company B). This requires active leadership and the entire workforce to be engaged. We must develop a culture where everyone believes that all incidents are preventable and that "zero incidents" is achievable. With engaged and committed leadership, effective processes and an OE culture, we can achieve our objectives in operational excellence.

This document provides an overview of the OEMS, our standard approach for achieving world class performance. It includes general guidance for the implementation and operation of the OEMS. More detailed guidance can be found on the Operational Excellence Website and within OE documentation, the Leader's Guide to the OEMS and in the online OE Certification modules.

Company C

• Company C describes their HSSE management system with:

To achieve continuous performance improvement (Company C) Companies manage health, safety, security, environment, and social performance in a systematic way. At (Company C), we aim to help meet the energy needs of society in ways that are economically, environmentally, and socially responsible. To manage the impact of our operations and projects on the environment and society we have a comprehensive set of business principles and rigorous standards covering health, safety, security, environment (HSSE), and social performance (SP).

Our business principles provide high-level guidance and the Commitment and Policy on HSSE and SP reflects our aims on how we operate and involve communities close to our operations. Those aims include:

- Do no harm to people;
- Protect the environment;
- Comply with all HSSE laws and regulations.

All (Company C) companies, contractors and joint ventures under our operational control must manage HSSE and social performance in line with the Commitment and Policy, local laws and the terms of relevant permits and approvals. To help our staff and contractors to put the Commitment and Policy into practice we launched the HSSE and SP Control Framework. It includes a set of mandatory standards and manuals covering areas such as managing greenhouse gas emissions, impacts on biodiversity, road safety and security. We also include requirements for integrating environmental and social factors into the way we plan, design and take investment decisions on new projects.

Company D

- Company D describes their operational management system with:
  The Operating Management System (OMS) framework, when fully implemented, helps to deliver safe, responsible and reliable (Company D) operating activity and continuously improve performance.

The OMS Framework

Every site and business within (Company D) currently has a management system to set priorities and manage risk. The purpose of the OMS framework is to help bring the appropriate level of consistency and completeness to all these systems.

OMS is a framework that defines a set of operating requirements. It sets out a systematic way to improve local business processes to deliver these requirements. When fully implemented, it helps to deliver safe, responsible and reliable (Company D) operating activity and continuously improve performance.

Safety Management

2

Keywords

Accident ratios; Accountability; Attitude; Behavior; Casual factor; Coaching; Communication; Cost of accidents; Culture; Deflecting; Discipline; Education; Elements; Employee involvement; Engagement; Ethics; Fishbone diagram; Iceberg model; Incident investigation; Integrity; Leadership; Leading by example; Legal; Lessons learned; Listening; Loss triangle; Near miss; Observations; Performance; Probing; Procedures; Reflecting; Root cause; Simultaneous operations; Skills; Training.

Operational Excellence finds its roots in safety management for the simple fact that safety management permeates every facet of an organization's operational efforts. Safety management has often been described as the control of recognized hazards to attain an acceptable level of risk. Simply put, to manage the business, you must manage safety effectively. Many understand that safety is key to achieving operating excellence (OE) goals and objectives and that OE broadly encompasses facets of the business that include provisions such as environmental management and asset integrity. While these may have their own specific areas of focus and be viewed as functionally different, it is clear that these are also very interrelated and that an organization's safety performance is key to achieving overall business goals and objectives. All too often in the oil, gas, and petrochemical industry, incidents that start out as being thought of as "safety" in nature morph into something more. For example, as we discussed with the BP Deepwater Horizon incident occurring on their offshore Macondo prospect, an incident that began as an operational abnormality quickly grew into something bigger and indeed became an incident with very material consequences. First, the inability to maintain positive control of the well led to a release of gas, which in turn led to an ignition of the gas and a serious explosion, leading in turn to injuries, fire, and loss of life, but with these very clear safety failures, the incident also then morphed into a very serious environmental disaster with uncontrolled release of crude oil into the Gulf of Mexico. So while the incident has its origins as a serious operational safety related incident it quickly escalated into something much greater in scope and impact. This is just one example of how interconnected safety is with just about all the other disciplines within an organization. Over the years we have heard many industry executives observe "how you manage safety is how you manage everything" and for those employed in the oil, gas, and petrochemical industries the need to manage safety effectively is obvious and underscored to get it right by the industry's high profile failures. Indeed, production, quality, cost, and loss control are of equal importance in measuring job performance and cannot really be separated; managing safety is integral and fundamental part of managing the entire business.

Applied Operational Excellence for the Oil, Gas, and Process Industries. http://dx.doi.org/10.1016/B978-0-12-802788-2.00002-6
Copyright © 2015 Elsevier Inc. All rights reserved.

Ethics and Safety Management

Many in management regard the issue of ethics as a simple matter of personal scruples, a confidential matter between individuals and their own consciences. And these same individuals are often quick to describe any wrongdoing as an isolated incident; the work of a misdirected (rogue) employee. The thought that the management of the company could bear any responsibility for an individual's misdeeds may never enter their minds. Hoping to distance themselves from responsibility and accountability for the resulting problems, they assure themselves that "these problems of ethics have nothing to do with management." The only problem with that line of thinking is that it is incorrect—ethics has *everything* to do with management. Rarely do character flaws of a lone individual fully explain corporate misconduct. More typically, unethical business practice involves the tacit, if not explicit, cooperation of others and reflects the attitudes, values, beliefs, decisions, language, and behavioral patterns that define an organization's operating culture. Ethics, then, is as much an organizational as a personal issue. Managers, who fail to provide proper leadership and to institute systems that facilitate ethical conduct share responsibility with those who conceive, execute, disguise, ignore, omit, and otherwise knowingly and deliberately benefit from corporate misdeeds—or even those that just don't care and choose to do the wrong thing.

In light of the outsized consequences for getting it wrong and having to deal with the tremendous cost of failure (in both monetary and human terms), many organizations are recognizing safety and environmental issues as key ethical issues. And in an effort to instill basic training on business ethics and excellence in the executive suite, many of the top tiered universities have begun in recent years to offer elective courses as part of their Master of Business Administration (MBA) curriculum, as well as continuing education encompassing operational excellence related coursework aimed at developing management skills for health, environment, and safety. For example, the Harvard School of Business offers a course called *"Why You Should Care: Creating the Conditions for Excellence"* (course number 2155 offered Winter session 2015) and *"The Moral Leader"* (course 1562 offered in the fall). They also provide through their Executive and Continuing Professional Education in the School of Public Health, coursework addressing transformational leadership training for achieving engagement and functional excellence. Similar course offerings are included in the curriculum at other leading universities such as Stanford's Graduate School of Business with first year course offerings such as *"Ethics in Management"* and *"Strategic Leadership"* (autumn quarter) which examines culture in shaping organizational performance.

As discussed earlier when reviewing corporate governance in the context of OE, strong OE leadership is important, because it shapes and molds an organization's corporate culture, and it is this culture which defines and influences employee behavior with respect to safety. Tasks involving safety may be delegated, but responsibility and accountability will always remain with the organization's senior leaders, so it is essential that they promote an environment which fosters and encourages safe behavior.

OE functions are typically structured to work together so as to assure operational discipline and this, coupled with other formal internal controls (i.e., financial accounting related), all serve to assure that the company effectively maintains its fiduciary responsibility to the company owners.

Senior managers must acknowledge their role in shaping organizational ethics and seize this opportunity to create a climate that can strengthen the relationships and reputations on which their companies' success depends. Executives who ignore ethics run the risk of personal and corporate liability in today's increasingly tough legal environment. In addition, at least in one locale, they deprive their organizations of the benefits available under new US federal guidelines for sentencing organizations convicted of wrongdoing. These sentencing guidelines recognize for the first time the organizational and managerial roots of unlawful conduct and base fines partly on the extent to which companies have taken steps to prevent that misconduct.

Prompted by the prospect of leniency, many companies have rushed to implement compliance-based ethics programs. Often designed by corporate counsel, the goal of these programs is to prevent, detect, and punish legal violations. But the entire organizational ethics issue means so much more than avoiding illegal practice; and simply providing employees with a rule book will do little to address the problems underlying unlawful conduct. To foster a climate that encourages exemplary behavior, corporations—*led by their senior management team and driven through the line management chain of command*—need to establish and maintain a comprehensive approach that goes beyond the very limited (and often punitive) legal compliance stance. This often involves a performance approach that advocates continuous improvement so as to apply lessons in a manner that builds upon successes and that learns from failures in a positive manner. And an Operational Excellence management system is just this sort of comprehensive effort many organizations use to shape and drive exemplary behaviors and decisions on a continuum by the workforce.

An integrity-based approach to ethics management combines a concern for the law with an emphasis on managerial responsibility for ethical behavior. Though integrity strategies may vary in design and scope, all strive to define companies' guiding values, aspirations, and patterns of decisions making as well as daily conduct. When integrated into the day-to-day operations of an organization, such strategies can help prevent damaging ethical lapses while tapping into the human desires for positive achievement and alignment with basic moral principles. Then an ethical framework becomes no longer a burdensome constraint within which companies must operate, but the governing ethos of an organization.

Errors of judgment rarely reflect an organizational culture and management philosophy that sets out to harm or deceive. More often, they reveal a culture that is insensitive or indifferent to ethical considerations or one that lacks effective organizational systems. By the same token, exemplary conduct usually reflects an organizational culture and philosophy that is infused with a sense of responsibility. As more senior managers enhance and expand their perceptions of the importance of organizational ethics, many ask their legal counsel to help develop corporate ethics programs that serve to detect and prevent violations of the law. These are often supplemented by formal establishment of corporate values and link to robust health, safety, environmental management systems, and engineering standards aimed at assuring asset integrity under an Operational Excellence framework.

From MBA Oath to Successful OE Executive Management Leadership

In the early morning hours of December 3, 1984, a poisonous gray cloud (comprised of some 40 tons of toxic gases) from Union Carbide India Limited (UCIL's) pesticide plant at Bhopal spread through the city. Water carrying catalytic material had entered the Methyl Isocyanate (MIC) storage tank number 610. What followed was a nightmare of epic proportions. The toxic gas spread through the city, sending scurrying residents through the dark streets. No alarm ever sounded a warning and likewise, no evacuation plan was prepared or executed. When victims arrived at hospitals breathless and blinded by their exposure to the toxic gas, attending physicians did not know how to treat them as UCIL had not prepared emergency information. When the sun rose the following morning, the extent of the devastation became painfully clear. Dead bodies of human beings and animals littered the streets, leaves changed color to black, and a smell resembling that of burning chili peppers permeated the morning air. Estimates of the death toll indicated that as many as 10,000 may have died immediately and some 30,000 to 50,000 were too ill to ever return to their jobs.

The catastrophe raised very serious ethical issues. The pesticide factory was built in the midst of densely populated settlements. UCIL chose to store and produce MIC, a very toxic and even deadly chemical (permitted exposure levels in the US and Britain are 0.02 parts per million), in an area where nearly 120,000 people lived. The MIC plant was not designed to handle a runaway reaction. When the uncontrolled reaction started, MIC was flowing through the scrubber (intended for neutralizing the MIC emissions) at more than 200 times the designed capacity.

The circumstances of this tragic incident were the topic of great discussion and debate for many years that followed. Not only was there the devastating human toll to address but there were years of legal wrangling to sort out, with the company—*once considered an industry leader and notably one of industry's bellwether companies with its inclusion in the Dow Jones Industrial Average (DJIA)*—eventually fading from view as it was subsequently purchased outright by Dow Chemical in 1999 for some $8.89 billion in stock. Not only was this a topic that garnered a great deal of worldwide press, but in matters related to business ethics, this incident served as a case study in many business management and engineering classrooms over the years that followed.

Incidents like this as well as other high profile business failures were to some extent responsible for top tier business schools offering classes in business ethics. Some of the topics related to the Bhopal incident that were discussed in business ethics classroom case studies would include and encompass discussion and learning with respect to some of the following issues:

- Ethical values and legal principles are usually closely related, but ethical obligations typically exceed legal duties. In some cases, the law mandates ethical conduct. Examples of the application of law or policy to ethics include employment law, federal regulations as well as industry codes and consensus standards.

- Though law often embodies ethical principles, law and ethics are not mutually exclusive. The law does not prohibit many acts that would be widely condemned as unethical. And the contrary is true as well. The law also prohibits acts that some would consider unethical. For example, lying or betraying the confidence of a friend is not illegal, but most people would consider it unethical. Yet, although we can understand and agree that speeding is illegal, we also know that many people do not have an ethical conflict with exceeding the speed limit. Do supervisors routinely look the other way when they see an employee violate a safety rule? Did they see problems but ignore them in Bhopal? Did senior management apply different values and standards to their operations which ultimately put different values on people's lives simply by virtue of their location on the globe and the extent of the protective controls mandated under different legal frameworks? Law is more than simply codifying ethical norms and the practical application of ethical standards goes beyond just meeting the minimums established in the law—some basis of it often simply involves doing the right things for the right reasons. When the law and regulations are silent on an issue, how does the court system establish what a reasonable man would do under such circumstances to assure the safety of plant, property and people?
- What should be the view of a host country vis-à-vis multi-national corporations and vice versa in terms of ethics?
- If you join a company manufacturing or distributing a dangerous chemical, what steps would you likely take to be on the right side of the law and of your own personal ethics?

With some 150,000 fresh MBAs being minted in graduate schools across the country each year, many of whom move into key management positions in leading companies, the idea of enhancing industry ethical decisions by addressing the topic in business school graduate programs has gained traction in recent years. With the emphasis being put on ethical decision making and leading in a socially responsible and sustainable manner, MBAs have begun taking an ethical oath as they embark on their careers in the business world. Following the medical profession's lead with the Hippocratic Oath, and the engineering professions code of ethics for engineers, the MBA's oath is a voluntary pledge for graduating MBAs and current MBAs to "create value responsibly and ethically."

Although the Harvard Business School Class of 2009 graduates may credit themselves with creating an internet website (http://www.MBAoath.org) to coordinate and communicate activities surrounding the launch of the Oath website, the history of the Oath does not begin there. From the Hippocratic Oath to Thunderbird's Oath of Honor and the Columbia Business School's Honor Code, professionals, students, and academics evoke kindred aspirations to align ethical values and good faith actions among those engaged in the management profession. Since then, working with dozens of top tier business schools, they have created a community of new and experienced MBAs committed to high standards of ethical and professional behavior as well upholding the letter and spirit of the law.

Now claiming a broad coalition of MBA students, graduates and advisors, representing over 250 schools from around the world, the organization partners with the Aspen Institute and the World Economic Forum.

Management teams embark on the Operational Excellence journey to address the complex challenge of achieving world-class performance results and avoiding incidents, injuries, business interruptions, and economic losses. Many of the oil, gas and petrochemical industry challenges continue to exist, but how they are dealt with effectively determines the success of individual organization's management teams. Reflecting on earlier experience with the energy industry and identifying basic causes of incidents can help industry learn and apply lessons so as to help avoid recurrence. A review of the American Petroleum Institute Safety Digest of Lessons Learned (API Publication 758, Section 2, Safety in Unit Operations) recounts details of 88 incidents that occurred from 1959 through 1978. Noting that more than half of these involved a fire or explosion or both, another twenty-five percent occurred during periods of startup, shutdown and turnaround. Categorizing the causes of the incidents provides insight into the kinds and the frequency of the problems confronting operators. The report cites the following causes and presented them in the order of importance:

Equipment Failure—*Twenty-Eight Percent*; these failures occurred when materials of construction were unsuitable, weld problems developed, or unanticipated stresses occurred. Instruments, pumps, electrical systems, valves, piping, and vessels can be involved and subsequently may require modifications to correct the problem. On occasion, flaws are exposed only through experience.

Human Error—*Twenty-Eight Percent;* human error is just as important in these incidents as is equipment failure. In the past, human error usually stemmed from a failure to follow instructions, often because established procedures were not enforced or because more refresher training was needed. However, it was obviously contrary to the intent of normal operating procedures and training to stifle the exercise of common sense.

Faulty Design—*Thirteen Percent;* during actual operations, design discrepancies turn out to be more critical than surmised during the design, expansion, or modification of the process/unit. It can serve as a good example of a unit telling its operators what ails it.

Inadequate Procedures—*Eleven Percent;* after an incident on a unit, it may become apparent that the operating procedures require expansion and revision. For example, the partial loss of a unit due to power failure may be followed by an unsuccessful attempt to recover it. Such an incident can reveal that the operators did not have sufficient knowledge or training to correctly handle the situation.

Insufficient Inspection—*Five Percent;* when critical equipment failures occur (such as in vessels or lines, particularly if the latter have dead ended sections), the scope of inspection must be expanded. In reported instances, lines that had just been inspected failed in a

section of the line that was not covered in the inspection. This type of failure emphasizes the need for more effective inspection and safety audit techniques.

Process Upsets—*Two Percent*; serious process upsets were not frequent in the incidents covered in this section, which supports the premise that, with other factors being equal, the refining and petrochemical industry has the knowledge to control its operations.

Education—*Thirteen Percent*; this category is separate and appears last because it does not represent a single specific incident. It is information that one refiner has gained through experience by using a procedure that he wants to share with his industry contemporaries in the interest of safety. It is a credit to the oil, gas, refining, and petrochemical industry that this means of industry wide communication is routinely made available and widely disseminated (i.e., through industry associations and professional associations), and greater use of it should be encouraged to help prevent operational incidents.

For executive management to manage such a broad array of complex variables, it is crucial that they provide effective leadership in each of these areas in a well-organized cohesive fashion so there are no errors or ethical breaches that result in incidents involving injury, property loss, or business interruption. The list should prompt recognition of the people leadership challenges across a broad spectrum of issues. While they all involve managing people as they properly execute their responsibilities, these leadership roles require the application of multiple skill sets; some aspects involve direct supervision of behaviors during the course of tasks related to operational safety, whereas others involve managing and overseeing key functions such as engineering design, construction, inspection, maintenance management, or procurement (e.g., topics central to asset integrity). And they all require clarity on issues related to ethical and prudent behavior.

Consistency and uniformity in adherence to rules and standards is critical to achieving operational discipline and avoiding incidents. At times the workforce will look closely at the signals they receive from company leadership and take their cues accordingly based on behavioral expectations (both real and perceived). The signals management sends will be interpreted by the workforce not just on the basis of their spoken words, but to a degree as well on the examples set by their own demonstrated leadership behaviors. Does management hold subordinates accountable for following established safety rules while at the same time flaunting them himself? In circumstances where the expectation is one that essentially conveys "do as I say, *but not as I do*" the worker finds the excuse and justification for *not* following the rules in the example set by the boss. The conflicting signals that arise in such circumstances can create internal stress and conflict as the worker tries to decide the proper (and ethical) behavior expected of him as he works to earn a living to support his family. And in those circumstances, where the perceived conflict can involve the perception of taking unauthorized shortcuts that compromise his safety, it also impacts morale and productivity. Supervisors that do not demonstrate genuine care and concern for their workers have little hope to connect with them on a personal level and risk alienating them to the point of impacting productivity. Workers ask themselves "If the boss doesn't care, why should I?" and for some this becomes an internal ethical struggle as they proceed to do as little as possible to get by. For example, if the manager were to send the signal that an operational imperative (e.g., "*just get it done…*") outweighs or is somehow more important than taking the time to do the job safely, the worker may interpret the situation as a signal to shortcut established safety rules for the sake of restoring production quickly.

While this may not be the manager's intent, there are times when he may willingly look the other way as the worker takes those shortcuts. Similarly, written procedures and robust engineering standards that signal a very low risk tolerance can be set aside by the individual willing to push the limits exhibiting a personal risk tolerance outside accepted norms and not well aligned with company standards. Over time these factors can result in *defacto* changes (i.e., significant alterations to established procedures, safe practices, precautions, and all manners of shortcuts) that become the accepted way of completing a task. And for a while it might even seem to work out—as long as nothing bad happens in the process. The manager or supervisor may even reinforce the actions by upholding the actions of the employee to get production up and running with no mention of a serious breach of safety protocol. Unfortunately, such circumstances can also result in an undesired incident resulting in injury, property loss or delays. For those who have a drive to excel, and always do the "appropriate action," this can be particularly frustrating. For the manager seeking to build long-term value through operational discipline this can end up being a long-term nightmare. Industry leaders intent on establishing or further improving their operational excellence efforts understand how critical senior management leadership is. The following summary compilation was adapted from a wide breadth of aligned guidance for senior management. It originates in numerous guidance documents and best practices from within industry and parallels leadership guidance developed by the UK Health and Safety Executive in *"Leadership for the major hazard industries."*

Achieving a positive operational excellence culture is based on sound ethical practices applied diligently and consistently. Senior management teams striving for world-class performance and outstanding safety and health performance records go beyond strictly focusing on regulatory compliance; they work to develop their own best practices as they establish and maintain performance standards. In the spirit of the MBA Oath and the code of ethics for engineers, the following is presented as a senior management pledge for making OE work in their organizations:

Ethical OE Leadership

Establish an Effective OE Culture

As senior managers, we recognize the importance of achieving a positive operational excellence culture and that excellence is defined by the diligent consistent efforts by everyone in the organization. An organization's company culture reflects the collective actions of every worker and in our role as leaders we recognize we must always:

- Recognize that the attitudes and decisions of senior managers are critical in providing strategic direction, establishing and abiding by corporate values and adjusting priorities of the organization accordingly. The attitudes and actions of every member of the management team will impact on the styles of behavior and priorities of those at all levels in the organizational hierarchy. Members of management will organize training as necessary and appropriate so that all members of the management team have a clear understanding of OE expectations and individual roles, responsibilities for achieving world-class OE performance across the organization.

- Lead by example. This means our role is not limited to simply directing work and monitoring compliance with rules and regulations. All members of management will act as leaders and facilitators: they continually encourage suggestions, motivate their staff and engage with the workforce to proactively identify and solve health and safety problems.
- Know the operations and stay informed on what is really happening, not what subordinates think we want to hear. We must know where there are problems and where things could go wrong. We must remain approachable so that all staff are able to candidly express their concerns with us. When they do, we respect their positions and work with them to find workable solutions.
- Actively engage the workforce and be alert to unsafe conditions and address hazards. This means we will take the time to address conflict and unusual situations by taking proper corrective actions to resolve deficiencies and actively demonstrate our personal commitment to obeying all applicable rules and standards. We will work safely and comply with the rules, and also encourage initiative and proactive efforts at improving health and safety. Employee behaviors and decisions are very much key to shaping the company's OE culture—defined as much or more by what we do as with what we say we will do. The management team will engage with them and encourage joint involvement and cooperative efforts of supervisors with employees in safety activities, wherever possible.
- Take effective action and uphold safety, health and environment equal in stature and importance to production. It will not be viewed as a separate function, but rather an integral part of productivity, competitiveness and profitability and that our health and safety risks are recognized as an inseparable part of our business risks.

Leading by Example

The senior management team will personalize safety by setting a good example and emphasizing the importance of OE through all demonstrated personal actions and decisions. In our leadership role we will actively promote OE and emphasize safety. We will always:

- Actively demonstrate genuine care and concern for all in our care. We do this by making it clear that exemplary OE performance is our aim and that we value the safety, health, and well-being of our workers, contractors, visitors, and members of the public.
- Include health and safety on the agenda of any board or management meeting and ensure the company routinely reports our health and safety performance as part of our commitment to corporate social responsibility. We will similarly encourage that health and safety is on the agenda of management meetings at all levels in the company.
- Actively extend accountability for OE performance throughout the chain of command and highlight safety performance as a key element of all performance reviews and through regular inspections and field observations. Managers and supervisors are in turn are accountable for the health and safety performance of their departments. Clear expectations are communicated and specific roles and responsibilities assigned and adequate resources provided to assure commitments are met. Members of the line management chain of command will not abdicate their responsibilities and will not tolerate "shortcuts" or abrogating safety rules and standards. Members of management will take effective action to assure that the corporate safety value is effective and that everyone understands and takes effective proactive efforts to assure their own personal safety and that of others.
- Measure our health and safety performance through useful and meaningful indicators. These measures compare our performance both internally over time, and also externally against others working with similar hazards.

- Set long-term goals for the control of major hazards and health and safety equally as for financial and production goals and have a plan to meet these. Every opportunity for learning is taken and used in our drive for continuous improvement.
- Meet with the workforce regularly and candidly discusses health and safety. Management actively encourages staff to raise health and safety concerns and issues by acknowledging them and providing specific and timely responses to suggestions made.
- Award contracts to companies who can demonstrate a good health and safety performance and who have a good understanding of the hazards they will encounter and the capability to properly mitigate them while working for us. Members of management will regularly meet leadership from our contractor to review their health and safety performance against our clearly defined expectations and to understand how their activities can impact on our health and safety performance.
- Fully investigate all incidents and near misses to identify the underlying causes and follow up on the agreed corrective actions. While we may acknowledge that people make mistakes, we do not accept accident investigation reports that identify "human error" as the sole cause of an incident.

Supporting the OE Management System

The senior management team will personally support and actively promote the OE management system and make its effectiveness a prime consideration in all business activities. In our leadership role we will collectively and individually:

1. Demonstrate that we understand where in our activities major accidents and incidents can occur and that suitable engineering/technical and human controls are in place. This is not at the expense of conventional health and safety issues, but we understand that the control of major hazards is a business priority.
2. Be confident that our staff are competent to carry out the tasks they are required to perform. Our competence management system identifies safety critical roles and tasks and these are routinely reviewed.
3. Know we have developed key performance indicators for major hazards and that process safety performance is monitored and reported against these parameters.
4. Review all pertinent facts and circumstances surrounding incidents and thoroughly investigate all incidents and promptly take appropriate corrective action. Incident investigation

procedure ensures we consider all issues, including human factors. It ensures immediate, as well as underlying management-related, causes are identified, without attributing blame, prompt corrective action is taken to prevent recurrence and lessons learned shared with others.

5. Facilitate communication and encourage and enable dialogue so people discuss health and safety in the context of OE. We know our managers encourage the staff to be involved in making health and safety decisions wherever possible. Anyone can, when they perceive the need, intervene in the work process to prevent hazardous working and suggest safer methods. They are fully supported by every member of line management in this approach.

6. Take positive steps to assure asset integrity throughout the asset's lifecycle. We continually assess the technical integrity of plant and equipment so it remains fit for service and is operated in accordance with proper engineering design, address feedback from operations, assure thorough hazard studies and competent risk assessment as well as high standards of construction. We have systems and individual processes that effectively deal with all these issues.

7. Assure the technical integrity of our existing plant and equipment through good maintenance plans and in carrying out maintenance to the highest standards. Management processes, programs and procedures reassure me that all these issues are under control and they are independently audited.

8. Manage change effectively (including both organizational and technical change). We are confident the systems on which we rely are up to date and subject to monitoring and review.

9. Comprehensively review the company's performance, based on all sources of information including accidents, high potential incidents, verification of results and monitoring of the important performance standards.

Engaging the Workforce

The senior management team will show genuine care and concern for the safety and well-being of our workforce. We make safety personal, for ourselves as individuals just as we would for every member of our workforce. Accordingly, as individual members of management this is our personal pledge:

1. I understand that successful businesses increasingly encourage active participation of the workforce in the management of health and safety. I will ensure that I am tapping the knowledge of how to do things better, more simply and more safely, that resides in the people who work for us. This includes our own workforce representatives as well as contractors.
2. I know that involving staff in the process of identifying and managing risks is a key aspect of managing health and safety successfully. I know that there are a number of ways of involving my employees in improving our management of health and safety, for example (but not exclusively) using safety representatives (unionized or non-unionized), safety committees and work councils. I will use the most appropriate methods for this organization.
3. I review our progress against agreed objectives at regular intervals and set performance measures. Performance improvement never ends. As a result, I will continually identify opportunities for improvement and develop and effectively implement an improvement plan.

Managing Safety

The starting point for achieving world-class safety with most of the industry's leading successful Operational Excellence systems begins with embracing the basic principle that all occupational injuries and illnesses can be prevented. Industry experience indicates that more than 90% of all injuries and illnesses in the workplace result from unsafe acts or lowered behavioral standards. Because most environmental failures result from safety failures, behaviors are also central to incidents resulting in environmental impact. It is important to note that incidents resulting in injury or illness involving improper employee behaviors often indicate failures in implementing specific processes and programs within an organization's safety management system (SMS).

Over their many years of working in the energy industry, the authors have been able to synthesize a simple seven point philosophy that underscores some of the most important concepts of effectively managing safety and by extension OE. These involve the following basic issues that are key to safety in the energy industry; it might seem a bit overly simplified when discussing the complexities and interconnected aspects of safety management, but it still seems like a great place to start:

- Prevention of all injuries
- Responsibility of all management
- Safeguarding of all facilities and equipment
- Training of all employees
- Efficiency and economy of business
- Safe conduct is a condition of employment
- Each task, the right way—*every time*

Looking beyond the philosophical outlook and putting them into context, effective line managers master essential functions as they establish and maintain standards to manage safety in their respective operations. These include—*responsibilities, standards, documentation, training, measurement, and effectively extending accountability*. Coincidentally, these are also commonly reflected in most safety management processes and are factors addressed by multiple process improvement methodologies.

Without giving it a lot of thought, some people will make personal judgments and support it with inferences that refer to safety only in the past tense. For example, "this *is* a safe operation because there *were* no injuries over the past one million man hours." And while that may be the extent of the interest for some, such perspectives might reveal an incomplete view of the idea of what safety means by simply holding the end result as the sole focus of the concept. When we examine the topic of safety management it's necessary to look beyond the end results and look at the efforts that went into achieving those results. What is the organization doing proactively in a specific area to accomplish their results and how is it being done? It is as much a qualitative view as it is quantitative.

While it is certainly possible to achieve safe outcomes without specifically addressing the precursors or antecedents necessary to achieve acceptable safety performance, it's important to realize that such outcomes under these circumstances can often be chalked up to a stroke of luck or simply be a matter of chance. Organizations that have a lot on the line don't leave such critical performance measures to chance— they actively manage their circumstances to ensure they achieve the desired outcome. This generally involves approaching everything from complex projects down to the smallest individual tasks in a structured and orderly fashion with a planning mindset and proactively making sure that all potential hazards are identified and the proper measures put in place to avoid undesired outcomes.

The most important component in all safety management programs is unquestionably that of diligent proactive efforts to identify hazards, assess risks, and take the necessary steps to properly mitigate them so as to effectively prevent incidents that may result in injury or property damage.

Extending Accountability for Safe Operations

Overall accountability for safe operations rests with the manager. The manager's challenge is to enlist everyone's support for his vision of achieving world-class performance. The importance of each individual to successfully reaching that goal cannot be overstated. Business and operating plans detailing objectives and strategies for reaching goals are tools that help the manager successfully achieve safe, reliable operations. Performance measurement is an important part of developing and implementing these plans. The primary accountability tool for employees is the annual performance appraisal. It is particularly useful when the results of the appraisal reflect each employee's diligent efforts to proactively work to identify and mitigate hazards in order to achieve safe operations. For contractors and their employees, the contractor evaluation process tied to bid awards is one of the most important accountability tools to influence safe behaviors.

Everyone in an organization has some degree of safety responsibility that they are expected to fulfill as they complete their work on a day-to-day basis. These responsibilities may differ somewhat based on the individual job assignment, but, will generally involve at a minimum, the expectation for completing work in a safe and professional manner. The line management organization establishes and maintains operating standards and extends accountability for safety down the chain of command with each successive level holding his subordinates accountable for safety performance in his respective area. This means that the supervisor has the responsibility for ensuring his respective employees conduct their work safely and this involves holding them accountable in turn, for adhering to basic safety rules, properly applying safe work practices, and properly implementing applicable procedures. The supervisor is held accountable for his efforts in managing safety performance and so on, up the chain of command. We talk about extending accountability in this manner but what exactly does it mean? We are all familiar with the concept of disciplining our children when they disobey the rules or do something demonstrating behavioral norms that are generally not tolerated by their parents. Organizations usually have some form of disciplinary procedures for addressing behavioral issues with employees and at times organizational disciplinary actions may be associated with disregarding or disobeying established safety rules. Many will understand the concept of extending accountability as holding a subordinate accountable for their actions and applying some sort of discipline as punishment for wrong doing. While responsibilities and accountability are closely related, the concept here for management extending accountability for safety goes far beyond the simple application of an employee disciplinary program.

Discipline—It's Much More Than Punishment

The term discipline has a number of meanings. The first one discussed above relating to inflicting punishment by way of correction and training is one aspect of what the term should suggest. The other definition for the word discipline suggests activity, exercise or a regimen that develops or improves a skill as through repetitive training (e.g., well-coordinated and precise in execution as typified by military-like discipline beginning in the planning and readiness stages and continuing through successful execution of operations). Both concepts of discipline are important in safety management and both are used to work together for the sake of assuring safe operations. One of the objectives in Operational Excellence is to assure operational discipline, which in the context of operations in the oil, gas and petrochemical industries should suggest achieving a state of order and control punctuated by having each task completed the right way—*every time* for the sake of consistently achieving optimal execution results and ultimately reaching and sustaining world-class performance. And to be clear, by definition this should clearly indicate an expectation for completing it safely. When management extends accountability down the chain of command after noting an undesirable or unsafe act, condition or an actual incident involving injury or property loss, it often involves someone other than the person whose decisions and actions did not meet expectations in a certain situation in a single or multiple circumstances. The person accountable is not necessarily the same person whose behavior(s) resulted in a mistake, omission, oversight, or infraction of rules. When management extends accountability for safety, management must reconcile the decisions and actions of the person(s) managing or supervising that are functionally responsible to make sure that subordinates wouldn't be in such a position to be able to make such a mistake or complete work improperly. In other words, accountability is largely about ensuring the measures of control are in place and fully functional so that undesired circumstances and events are avoided through effective loss prevention efforts.

We have discussed the connection between incidents occurring and the potential for disciplinary action (punishment) after the fact, but the important point to make here is that perhaps the most important part of management's efforts for extending accountability will take place *before* an incident occurs. By extending accountability for safety and addressing unsafe acts or unacceptable conditions *before* an incident occurs and not waiting to address it *after* the fact, responsible management teams assure that they are tackling their safety responsibilities proactively rather than reactively. And extending accountability has got to be something more than lip service. It means taking the time and making the effort to ensure employee decisions and behaviors are aligned with standards and expectations and operations are running properly at all times. Once standards are established, as in a safety manual, safety handbook, operating procedures, operating limits, basic safety rules, management process, best practices, instructions, etc., the management team will continually work to ensure the standards are understood and maintained. There are numerous ways to ensure standards are understood, some of these include efforts such as employee orientation, safety meetings, mentoring, and coaching, as well as specific measures to assure standards are maintained including behavioral observation processes, facility inspection processes, audits, reviews and Job Safety Analysis (JSAs). Other ways to assure standards are understood include design reviews during front end engineering design, design basis scoping paper reviews, and interim project design reviews at various stages of the project design. Similarly, engaging and reviewing major capital project contractor's safety management plans and providing period onsite inspection and review also provide the management teams with information and feedback to validate whether standards are understood and being properly maintained.

High Reliability Organizations (HROs)

A high reliability organization has been defined as one that produces product relatively error-free over a long period of time. Two key attributes of high reliability organizations are that they:

- Have a chronic sense of unease, i.e., they lack any sense of complacency. For example, they do not assume that because they have not had an incident for 10 years, one won't happen imminently.
- Make strong responses to weak signals, i.e., they set their threshold for intervening very low. If something does not seem right, they are very likely to stop operations and investigate. This means they accept a much higher level of false alarms than is common in the process industries.

Source: Business Case for Effective Process Safety; OECD Environment, Health and Safety Chemical Accidents Programme, June, 2012.

If one were to ask a manager or supervisor when they should be doing safety inspections, if the answer is once a week or once a month, then it may be that they don't understand their own personal responsibilities. While it is important to block time out of one's schedule for the sake of formally completing such inspections, observations or reviews, each time a manager or supervisor walks through his area he should be conducting an informal inspection even if the primary purpose is only to check attendance or to determine whether materials and supplies are adequate.

When a manager notices that something is not functioning properly and he inquires as to the reasons for the discrepancy, a subordinate supervisor might respond to the observation by noting that he told the employee responsible to correct the situation and do better next time. That may not be sufficient if it is not backed up by more concrete (visible) supervisory efforts to ensure the proper corrective actions are put in place (e.g., allocation and mobilization of resources (personnel, capital, equipment, maintenance etc.), frequent inspections or behavioral observations of procedures and positive efforts to correct deficiencies). If an item is brushed aside or otherwise not addressed, it can create a situation of ambiguity. It can send the signal that the problem is of no concern, or otherwise create the impression that management has tacitly approved something less than what was established as a standard and even lead to the belief by the workforce that following that particular standard is really only optional. In other words it serves to lower or diminish the standard and this can in turn lower the performance. Thinking about this in the context of everyday operational situations, it is easy to see why management leadership is so important and the efforts for extending accountability crucial to the continued success of the organization.

Many managers have thought "What is the most effective way to hold other people accountable when I'm not getting the results expected of them?" This is one of the fundamental challenges of management effectiveness so it is not surprising that this sort of question would be raised.

The results desired rarely, if ever, just automatically materialize because the manager expects the best from his people. Desired results are only achieved by (1) clearly defining what the results are and what the end state looks like so everyone involved has a clear idea of what is targeted, (2) ensuring that everyone is involved and has the means to becomes genuinely aligned and willingly assumes accountability for delivering specified results, (3) actively and frequently monitoring progress against each of the results, (4) clarifying and redirecting as necessary to make the needed course corrections along the way, and (5) seeing it through to the end by staying actively engaged until all of the desired results are realized. These may seem fairly straightforward, but in fact they may involve difficult conversations, necessitate greater collaboration, and result in employees who feel more empowered and made a part of the decision making process. Because of how important these steps are to extending accountability, a results-oriented manager will find they are definitely worth the time and effort to put into use to help assure the right results.

So what if all these things are being applied properly and the results expected are still not being realized? First, it's important to reconfirm that this is really all being done correctly and that the manager is tackling this in the proper sequence; establishes standards, communicates expectations, assures proper alignment (e.g., resource availability, eliminate conflicting priorities, etc.) and inspects/observes/intervenes as necessary to keep everyone on track to reach the objectives. If this accountability sequence is firmly in place and the desired results still don't develop, find out why by holding candid accountability conversations with the person or group who is the critical link, i.e., the person (or group) who has the greatest potential for not delivering on his or her key task. Such conversations may start by simply asking this person, "What are the obstacles that prevent you from making the progress you really want to make on this project?" "What else can you do?" "What actions are you going to take and by when?" "What can I do to help?" The objective is to get an agreement to do what needs to be done and make sure that they are in a position to see it through to completion.

Having such a conversation with people is important because it helps them realize what obstacles are frustrating their efforts, gets them to identify the ones that they can influence, helps focus and enable them to figure out a way to overcome these obstacles, and finally makes sure the they have a plan, the resources and motivation and are focused on achieving it.

Here's an example of how this type of conversation has been used in action. Based on a lack of acceptable progress, the VP of operations became frustrated with an apparent lack of concern for an impasse on solving a process design problem that involved younger members of his engineering staff. In taking the time to hold an accountability conversation with several of the engineers on the design team, he asked them why they weren't making progress with the process design problems. He discovered that the obstacles they were facing prevented them from moving forward and this was simply based on their perception that the operations leadership was resistant to their new ideas. Once everyone properly focused on solving their process design problem, the team comprised of primarily junior staff engineers was deliberately altered to now include a senior operations mentor. This new team solved the problem and the resulting process design solution saved the company a lot of money in operating costs. Once the motivation problem along with the associated company culture barriers were addressed, the VP's desired results and his vision for the end result was reached. This proved so effective that this senior manager implemented the approach of deploying teams comprised of junior staff engineers and senior operating mentors to attack other issues subsequently with great success. Many of the leading petrochemical companies have now harnessed the added value of mentoring by senior staff. So-called "short service employee" (SSE) programs are also used to pair very junior staff with more senior staff with the idea that they mentor them by providing guidance and instruction along the way and watching over them as they learn what is expected of them by their employer.

Training Employees

ANSI Standard ANSI/AIHA/ASSE Z-10, Occupational Health and Safety Management System (OSHMS), establishes the standards for organizations to "ensure through appropriate education, training or other methods, that employees and contractors are aware of applicable OHSMS requirements and their importance and are competent to carry out their responsibilities as defined by the OHSMS." The subjects which are recommended are safety design, incident investigation, good safety practices, hazard identification, and use of personal protective equipment (PPE). We have put a great deal of emphasis on one of the most important characteristics of an effective OE program—*leadership*—and effective training is but one factor among many that are necessary for employee training to be useful and effective. Training is not a substitute for supervision, and this means to suggest that it is not enough to simply send an employee to training and then expect flawless execution with no provisions for supervisory oversight. This notion relates back to the concept just mentioned for extending accountability and providing sufficient supervisory oversight to assure operational discipline across all operations.

The starting point for world-class safety is the principle that all occupational injuries and illnesses can be prevented. Industry experience indicates that more than 90% of all injuries and illnesses in the workplace result from unsafe acts or lowered behavioral standards. Orientation, training, and communication are key for an individual to understand what is expected of him by his supervisors and to properly equip the worker to perform his duties and responsibilities and supervisory oversight is important because it serves to establish and maintain standards. How much and what sort of training is necessary? How much is enough? How much is too much? How much time must the supervisor spend to make sure standards are maintained? How much and what type of communication is necessary? There is not a one size fits all answer to these questions, because each situation is different. It is sufficient to say that these functions are interconnected and one relies to a degree on the effectiveness of the others. We can tell you that many of the leading companies in the energy industry have invested considerable resources in ensuring members of their workforce are well trained, equipped, and qualified for their respective roles. Beginning with the educational and experience qualifications necessary for a new hire employee to be considered for employment to the extent he is further trained and developed to be a productive member of the workforce, most organizations spend considerable time and effort to make sure workers are properly equipped to handle their assignments and contribute to successful, sustainable, and injury-free operations. Most organizations develop succession plans to assure individuals move onto progressively more responsible positions with the benefit of institutional knowledge, education, experience, skills proficiency, and a proper mindset coupled with a record of solid achievements behind them. Some organizations develop competency maps defining technical skills and training necessary to help supervisors and managers plan work assignments and career development to ensure each person is well qualified and prepared for the roles to which he is assigned. Some organizations train and develop their workers through an individualized mentoring process, coupling more experienced hands with junior employees for the sake of knowledge transfer and on the job training.

Training, Behaviors and Performance

Training alone won't ensure that each and every decision and behavior of the employee will be correct and flawlessly executed. A company's engineering standards reflect its risk tolerance from a design engineering perspective, the company's operating procedures reflect its operating risk tolerance, maintenance procedure and equipment manufacturer's operating instructions reflect necessary precautions and mitigation to avoid incident. It is important to highlight these interrelated mechanisms, because it quickly becomes obvious that an employee's decisions and actions are key to assuring each task is completed the right way—every time. Just as the design of a facility to a specific standard requires it be constructed to a certain specification, the way the facilities are operated and maintained per operating and maintenance standards are also key. An employee who demonstrates a personal risk tolerance that is not aligned with these can pose a very serious risk to operational discipline. Uncorrected it can lead to very serious consequences. Let's examine a very simple example; you can have the best engineered and constructed set of stairs in a facility, but, if they are not coupled with the necessary employee behaviors of holding the handrail while ascending or descending the stairway, a serious potential for injury can exist should the user miss a step or unsupported by the lack of a firm handhold on the handrail or railing he may stumble and fall while using the stairs. Why is such a trivial sounding behavior so important? First, failure to properly apply these behaviors results in many slip, trip, or fall injury incidents across industry each year of which many could easily be prevented through the simply application of a basic safety rule. Second, if the individual is not willing to accept and positively demonstrate adherence to a basic simple rule like holding the handrail throughout his daily routine, what are the chances he will ignore those that present more of an inconvenience to him? How difficult do you think it will be for the supervisor to assure adherence to other more complicated (and perhaps cumbersome or otherwise inconvenient) rules and procedures? Many organizations present this very simple straightforward basic safety rule in an employee safety handbook because slips, trips, and falls can just as easily injure and employee as an operational safety related incident (and in some cases even more easily as walking is considered to be something so routine). Furthermore, an employee's willing acceptance of simple basic safety rules can be an important indicator for a supervisor to accurately gage and predict his willingness and reliability with respect to adhering to other more complicated rules and procedures.

A facility that has been properly designed, constructed, maintained, and operated will typically be operated safely within established process design limits (operating envelope) and operating outside these limits may result in undesired consequences (e.g., if a worker misses or ignores a high level alarm and takes no action or the incorrect action). Organizations with strong safety cultures, characterized by what the workforce does as opposed to just what they say they will do, benefit from greater operational discipline and efficiencies. With the considerable overlap in management controls (e.g., standards such as those associated with asset design, construction, operation and decommissioning, basic safety rules, procedures, training, communication, etc.) the manner in which these are managed will have a certain degree of impact on the others, in the efforts to properly control hazards.

Whatever an employee's motivations, if for some reason he chooses to make decisions or demonstrate behaviors that are at odds with formal training, established rules and standards and takes an improper course of action, fails to take the right action at the right time (effectively lowering a standard) or otherwise deviates from the proper behavior (e.g., shortcuts), this can lead to an unsafe act or condition. The people within an organization may distance themselves those involved after an incident by describing an employee as being "rogue" or acting improperly, or not following procedures or by using any number of similar labels. But as explained earlier in the section **Extending Accountability for Safe Operations**, an effectively run operation is a line manager's responsibility and these sorts of circumstances are an important part of his duties is to oversee and assure the safety of all in his care; he would not tolerate unsafe acts or conditions to occur or continue. A supervisor staying apprised of the work being conducted would provide the oversight to correct such circumstances. Some companies have also granted *stop work authority* to all members of the workforce with the clear expectation that if they see it "they own it" and they are clearly empowered to step in and intervene to correct the situation, and if necessary stop the work until the situation is resolved safely. Employees taking incorrect, improper actions demonstrating inappropriate behaviors or making poor decisions are to be immediately corrected or otherwise redirected by supervision to maintain order and discipline in facility operations. Left uncorrected, a rogue or misdirected employee taking shortcuts or deliberately circumventing established rules and procedures can be problematic and can quickly undermine an organization's operational discipline. If the situation is left uncorrected by supervision, it may appear to give a tacit approval to the unacceptable behaviors and this can allow things to escalate as others copy, mimic what is otherwise considered an unacceptable lowering of a standard established by authority.

Training and supervision are both important ideas and these functions, done properly one will complement the other. Under the right circumstances where the sum total of all mitigation measures, are not sufficient to control a hazard, injury and/or losses may occur. Incidents resulting in injury or illness involving improper employee behaviors often indicate failures in implementing specific processes and programs within an organization's SMS. Many companies in high-hazard sectors have made important strides in establishing the necessary corporate culture and leadership to minimize the frequency and severity of process safety and occupational safety related incidents.

Procedures

Written procedures typically include details of safety and health considerations (e.g., such as properties of and hazards presented by chemicals used). They will include precautions necessary to prevent harmful exposure, including engineering controls, administrative controls, PPE, and control measures to be taken if physical contact or airborne exposure occurs. They also identify special or unique hazards as well as detail safety systems and related functions. Operating procedures must be easily understood by and readily accessible to personnel who work with or maintain plant and equipment. They are reviewed as often as necessary (usually on an annual cycle) to ensure that they accurately reflect current operating practice. Changes to operating facilities potentially affect operating procedures. Such changes are subject to the management of change (MOC) process. Many organizations will opt to review and update their procedures when such changes affect hardware or process changes (which in turn affect operating procedures).

Operating Parameters

It is important to ensure that equipment is operated within safe operating limits throughout its entire service life. Management teams begin this by arranging to gather and compile data and all relevant operating limits. Information used to develop these operating limits may come from several sources, including design documentation or any subsequent MOC data. Operating limits are normally included in the operating procedures and operators made aware of them as well as the consequences of deviation along with the steps required to correct or avoid deviation and maintain safe and reliable operations. It is customary for appropriate alarms to be established within the control systems and covered in operating procedures. Factors related to production increases and operational upsets can have an impact and can result in exceeding safe operating limits. When established by authority, operating limits and set points should be viewed as a minimum until changes are authorized through the MOC process.

Equipment Startup and Shutdown

Losses associated with startup and shutdown of equipment and entire units are common (e.g., a notable industry incident occurred in 1989 with the Piper Alpha North Sea platform loss). As with any operating procedure development, employees should be involved for their insight and concurrence. Particular care should always be given when developing equipment startup and shutdown procedures. This includes commissioning new facilities (e.g., purging and leak testing).

Simultaneous Operations

Operating and organizational procedures will be developed and well-coordinated and communicated before beginning simultaneous operations that cross organizational boundaries.

These procedures ensure proper coordination of operational plans, including potential shutdown and timing for startup. These procedures address the transfer of control as much as they provide a basic delineation of responsibilities and actions to take, especially in emergencies. Examples of simultaneous operations include:

- Offshore marine operations involving multiple vessels working in close proximity
- T&Is amidst other operating units
- Drilling and work-over on producing platforms
- Construction within operating refinery units
- Pipeline construction or maintenance adjacent to a gas plant

Communications

Effective communication is a two-way process. Good communicators express themselves clearly, but also need to listen to what other individuals have to say and to give their messages thoughtful consideration. Effective communication between management and employees will assist in the organization functioning as a holistic team which will avoid or reduce risks which completing tasks safely, effectively and efficiently. Candid communication with employees about the company's objectives, risks, hazards, safety performance, and other concerns, helps to enhance understanding, motivate the overall workforce, instill employee confidence in management decisions and encourage employee involvement in the company operations. Having employees, supervisors and managers with good interpersonal skills such as collaboration, communication and conflict resolution allows strong relationships to development and which also allows individuals to become more competent, capable and confident in their abilities and to reach for higher aspirations, which overall, is good for the organization.

Communications Plan

Sharing information about the organization requires clear, direct communication with both internal and external audiences. Successful two-way communication involves exchanging information in a way that it verifies that it has been sent, received, and mutually understood. There could be instances where a one-way communication may be quite effective as well, such as use of posters, magazines, lectures, presentations, and intra- and internet websites.

A communications plan will provide details of formal and informal communications. It will also define appropriate communication interfaces that are necessary. Communication interfaces are those point of time, e.g., people, groups, locations, when the communication occurs, which may be internal or external. The audience, timeliness, importance, accuracy, and sensitivity are all factors which must be considered when determining what information needs to be communicated.

A communications plan will typically address the following:

- What information is to be communicated
- Who is to receive the information
- Who is to communicate the information
- How the information is to be communicated (e.g., formal, informal, etc.)
- When the information is to be communicated (e.g., real time, periodically, etc.)
- What measures are to be utilized to determine how effectively the information has been communicated.

Effective Listening

Research has found that by listening effectively, you will get more information from the people you manage, you will increase others' trust in you, you will reduce conflict, you will better understand how to motivate others, and you will inspire a higher level of commitment in the people you manage. Almost everyone sincerely believes that he or she listens effectively. Consequently, very few people think they need to develop their listening skills. But, in fact, listening effectively is something that very few of us can do. It's not because listening effectively is so difficult. Most of us have just never developed the habits that would make us effective listeners.

You Probably Don't Listen as Effectively as You Think You Do, and You Probably Don't Know It

A study of over 8000 people employed in businesses, hospitals, universities, the military and government agencies by researchers found that virtually all of the respondents believed that they communicate as effectively or more effectively than their co-workers. (Could everyone be above average?) However, research shows that the average person listens at only about 25% efficiency. While most people agree that listening effectively is a very important skill, most people don't feel a strong need to improve their own skill level.

Why Effective Listening Matters

To a large degree, effective leadership is effective listening. A research study of managers and employees of a large hospital system found that listening explained 40% of the variance in leadership. That's a big correlation by social science standards.

Effective listening is a way of showing concern for subordinates, and that fosters cohesive bonds, commitment, and trust. Effective listening tends to reduce the frequency of interpersonal conflict and increases the likelihood that when conflicts emerge they will be resolved with a "win–win" solution. In addition, if you listen to the people you manage, you will learn "what makes them tick." When you know what makes them tick, you will be more effective at motivating them. You can encourage them when they need encouraging, and you will know what kinds of things they value as rewards for a job well done (e.g., additional responsibility, public praise, autonomy, etc.).

What Is Effective Listening?

Effective listening is actively absorbing the information given to you by a speaker, showing that you are listening and interested, and providing feedback to the speaker so that he or she knows the message was received. Delivering verbal communication, like writing a newsletter, involves trying to choose the right words and nonverbal cues to convey a message that will be interpreted in the way that you intend. Effective listeners show speakers that they have been heard and understood.

How the Most Skilled Communicators Respond when Listening

The most skilled communicators match their responses to the situation. In discussions with the people you manage, it helps to differentiate the coaching situations from the counseling situations. Coaching is providing advice and information or setting standards to help your employees to improve their skills and their performance. Counseling is helping subordinates recognize and address problems involving their emotions, attitudes, motivation, or personalities.

The most common mismatch of response types to situations is the tendency a lot of us have is to give advice or deflect in a situation where counseling is appropriate. When you are counseling, "reflecting" and "probing" are usually more appropriate responses than "advising" or "deflecting."

Reflecting

As mentioned above, when we listen we should show the other party that what they are saying to us is being heard. Since we can think at about four times the speed that speakers can speak, our brains have a lot of capacity that can be used to process the meaning of what's being said. Reflecting is paraphrasing back to the speaker what they said. One of the things a lot of us find when we try to use this technique is that it's real a challenge! We don't want to just parrot back what was said; we want to paraphrase. It takes creativity to think of appropriate ways to paraphrase what we've heard.

Reflecting can take other forms than paraphrasing back to someone what was just said. For instance, a listener can summarize what he or she heard and also take the conversation a step further by asking a question for clarification or elaboration. One of the things we often notice when we reflect during a conversation is that the meaning we have ascribed to what we've heard has missed the speaker's intended meaning. When speakers hear us reflect, they get a chance to correct any misunderstanding that we have.

For most of us, it takes a lot of practice before we become natural and effective at reflecting. Our first few efforts may sound forced, phony, patronizing, or even "moronic." However, that doesn't mean we should give up learning how to reflect.

Probing

In addition to reflecting, the most skilled communicators' responses in counseling situations involve a lot of probing. Probing means asking for additional information or clarifications. Not all questions you might ask will be effective. Avoid questions that challenge what has been said because that will put the speaker on the defensive (e.g., "How could you have thought that?"). In addition, a question that changes the subject before the current subject is resolved isn't effective communication. Effective probing is non-judgmental and flows from what was previously said. Good probing questions ask for elaboration, clarification, and repetition (if, for instance, an important question you asked wasn't answered).

Deflecting

Deflecting responses shift the discussion to another topic. When we deflect from what we've been told, rather than acknowledging it, we can unintentionally communicate that we haven't listened and that we aren't interested. Deflecting shows that we're preoccupied with another topic or in some instances deftly avoiding an issue to be less than forthright or even deceptive in controlling a conversation.

Many of us deflect unwittingly by sharing our personal experiences when we should be focusing on the other party. Think about this from the speaker's perspective: You don't feel like you've been heard when you share a concern with someone and they respond by telling you about themselves. The responder gives you the impression that they aren't even listening, and that they just want to talk about themselves. Sometimes we mention our own experiences as a way of saying that we can relate to the speaker's experiences. Our intention is to say, "You are not alone." But, when we tell our stories we risk sending a message that we aren't listening and really don't care about the other individuals. Don't be a "Topper"—the kind of person who can tell a story to top any story that they're told. We all know a Topper, don't we? In a small way, toppers communicate that they are *superior*. That's not supportive!

This is not to say that sharing your experiences is never a helpful. On the contrary, mentors often help their protégés by relating their own experiences as a way to reassure their protégés that their concerns are normal and that their problems are solvable. But, in counseling situations, be careful to use deflecting only at appropriate times.

Speakers may not know that you have heard and understood what they have said if you deflect by moving on to another topic or shifting the focus to yourself or your own experiences.

Advising

It can be insulting to give advice to someone who has shared his or her problems with you. Sometimes individuals at work ask, "How are things going?" On the days when you would groan about a problem, a co-workers response would be to suggest what I should do about it. That may bother some individuals. They may value self-reliance and like solving puzzles, and may not really like someone telling me how to solve their problems. They may feel that the supervisor or manager didn't respect my ability to solve my own problems. They may want the self-satisfaction for finding the solutions by themselves, and want those individuals to respect their problem-solving abilities. So this type of communication style doesn't support that. I'm sure the advice such a manager gave was well intended. Nevertheless, they may not want to hear it. In fact they may ask for advice. Some prefer just to have the supervisor to listen instead of advising, they would have shared more and would probably have built a stronger bond. Instead, such advising caused the individual to clam up and it undermined his ability to understand what the person was going through. It has been found that this problem is particularly common between men and women in the workplace.

Women often discuss their problems and concerns with men just as a means of developing interpersonal bonds. When men respond by giving advice, they may believe they are being helpful to their female counterparts. But, if no advice is solicited then providing it is a little presumptuous, and it actually undermines the opportunity to further develop a cohesive bond with that female coworker.

Sometimes you might have a real problem keeping your advice to yourself. When someone is telling you about a problem they're having, you might barely control the impulse to tell them what they should do. You should be aware that people usually don't want your advice. When you give unsolicited advice, you have stopped listening and started to dominate the dialogue. (Imagine how frustrating that can be for employees.)

If you like to give advice, try fighting the urge as long as you can. Just reflect what you've hear and probe for additional information. Then, when you think the time is right to provide your words of wisdom, say something like, "Let me know if you'd like some advice. I've got some thoughts about that." You might be surprised by how few people take you up on that offer.

Typical Objections to These Effective Listening Techniques

Here are three common objections:

Reflecting slows down the conversation and wastes time. Yes, your time is a valuable resource, and you do want to invest it carefully. Reflecting takes time, but it can save time too. Many times reflecting does more than show the other party that they are being heard; it also serves as a check for accurate understanding and provides an opportunity for clarification. Reflecting takes time, but correcting errors due to miscommunication will be more complicated and time-consuming.

Reflecting might seem phony, patronizing, or moronic. Skilled listeners know that tactfully showing that you have heard what someone has said by reflecting it back to them requires creativity, and they've had to practice creative paraphrasing and reflecting to become good at it. Yes, the process of learning how to use reflecting can be awkward for people who are inexperienced with it. However, be very careful not to avoid practicing and learning a skill just because you're concerned that you will not immediately be proficient. It's better to develop communication skills over time, despite the possible awkward stage, than to completely avoid developing those skills due to a fear of the initial awkwardness.

"I don't have time to be the confidante of all my direct reports." Yes, there is a time-management issue. It might seem that the best way to use your time is to hear the problems, give advice, and move on. That may or may not be good time management. Think carefully about the consequences of showing your staff that spending time listening to them is not important enough to be a high priority for you. Managers who make it a high priority develop strong relationships, employee commitment and a support network for themselves.

Practicing This Management Skill

To develop your listening skills, plan to use the response type that you think you need to emphasize (e.g., reflecting) and plan to avoid using the response types that you want to de-emphasize (e.g., advising). Then, after you have a conversation, evaluate how effective you were at giving good responses as a listener. Identify what went well and where the opportunities for improvement are. Think about what that challenges to being an effective listener were and how you can deal with those challenges more effectively next time.

> *Good things happen when we integrate sustainability into our products, services and solutions. We improve our competitiveness and create and capture customer value. We save money, reduce our environmental impact and improve our employee satisfaction.*
> *Chairman and CEO Caterpillar—Jim Owens*

Monday mornings are usually a good time to practice your effective listening. Just start a conversation with a coworker by saying, "How was your weekend?" From there, just probe and reflect. In 10 min, you can actually get to know the other person a little better and show that you're interested in them. You can then utilize this approach in more important business conversations.

Making a recording of a conversation, if you can find a willing partner, can also help you evaluate your performance. With a recording of a conversation, you can examine each response you give in detail, without relying on your memory.

Communication, Coordination and Employee Involvement

All the points raised above are factors that influence how well an individual communicates with others at work (and off the job as well). Understanding and practicing the skills necessary to be a good communicator are important and can make a big difference in management effectiveness. Another equally important point is to make sure that the manager is taking the time to focus on communicating the right things at the right times.

Although the line manager has overall accountability for the effort, it is still the individual employee and their individual and the collective behaviors of all workers that determine the performance outcome. And while the actions of the workforce can have localized impact on those located in close proximity to those conducting the work, there are also circumstances where those located further away can be affected as well, such as when an uncontrolled release moves offsite to affect those adjacent to the site. For this reason, managers work accordingly to communicate loss prevention issues with all affected parties including management, employees, contractors, customers, suppliers, government agencies, and the public.

Managers and supervisors routinely involve employees whenever possible in planning, developing procedures, observing, reporting, and communicating. Their input and buy-in is often important to getting the job done correctly and because they are usually experienced and familiar with the site, equipment and procedures their insights and observations can be very helpful in communicating and coordinating effectively. Managers will typically devise a communications plan to address those critical points where direct communications are incorporated into daily operations. For example, transfer of operations at shift changes, simultaneous operations, and formal transfers from operations to maintenance and/or from company to contractors (and vice versa) are all points where communication interfaces are necessary for communication and coordination of information to effectively prevent losses. Management's plan for communication at these junctures will often be incorporated into procedural requirements (or otherwise incorporated into employee expectations so they are formally conveyed and acted upon) so as to routinely prompt communication at such important points during operations.

In addition, other venues for safety communication may be structured as well. Some examples may include establishment of safety committees and other safety forums can be used to exchange ideas on what works and what doesn't. Innovations and suggestions for improvements that help to encourage ideas for improvement should be actively sought and promptly addressed.

Employers are required, under health and safety regulations, and their own internal policies and procedures, to keep employees and their representatives informed about relevant safety and health issues—two such examples involve establishing the means for hazard communication and provision of process safety information to employees.

Employers are also required to train and educate employees on safety and health effects of chemical to which they may be exposed to, provide process safety information (e.g., hazards of chemicals used in a process, technology of the process, and equipment in the process) and to inform affected employees of the findings from incident investigations. A well-structured and organized OE management system will integrate these in such a way that the information is properly communicated and the proper corrective actions are promptly taken to prevent recurrence. Lessons learned can only be applied if they are communicated and others learn from what went wrong. Lessons learned can come from several sources, of course starting with incident investigation as mentioned, but also from others such as safety bulletins, near miss reports and employee experiences and observations. The task of getting this across to the right people can involve some effort by the line organization, and the importance cannot be understated—but let's not forget that communicating and coordinating is a very important aspect of effectively managing an organization.

Safety Inspections

Formal and information inspections are normally conducted by the line manager and if done in conjunction with formation of an inspection team will often involve other members of the workforce. Checklists are often used to prompt and remind of specific details to be aware of and look for through the course of an inspection. Managers typically conduct frequent inspections to help promote safe operations. Although inspections of general work conditions and facilities tend to target unsafe conditions (e.g., facilities equipment, hardware) more so than focusing directly on workers' behaviors, they are a valuable tool in safe operations. Housekeeping observations during an inspection provide insight into the overall standards set and maintained within a specific work location. Upon completion of an inspection a report is typically generated with findings and corrective actions documented and communicated as well as tracked through to completion and closure.

Behavioral Observations

Observation can be a very valuable management tool. Applied properly it can add value to many aspects of the business in addition to safety. The manager's safety observation tour provides the means to visibly demonstrate interest and concern for how work is performed and how the facility and equipment are being maintained. To effectively apply behavioral observation technique, line managers and employees will typically first be trained as skilled observers to distinguish between safe and unsafe acts. Being able to differentiate between safe and unsafe behaviors and being able to apply situational context to familiar settings can help the observer spot evaporative acts that might otherwise go unnoticed. An example of an evaporative act would be a situation where a worker is using a bench grinder without wearing the proper eye protection, but as soon as he notices the manager coming through the shop doing a behavioral observation tour he turns his head for a moment to quickly (hoping he is not caught in the act) put on the required safety glasses so he can avoid admonishment or disciplinary action. In such circumstances the worker knows he has done something wrong by not following a basic safety rule and the manager observing him can see from the worker's actions that he knows he is wrong in his veiled attempts to hide the infraction and correct it hopefully before he is held accountable for it. Managers should also be trained in the proper actions to take immediately after an observation and instructed on follow-up corrective action(s), including commending and reinforcing proper behaviors and correcting unacceptable ones. Typically, such observation tours take less than 30 min a day and should be scheduled events, actively involving other employees whenever possible. Failure to enforce a simple safety rule sends a negative signal to the employee, indicating indifference or neglect.

Observation/Review of Operating Procedures

Behaviors of employees and contractors applying operating procedures are often observed as well as and reviewed regularly. An example of an operating procedure that may be observed might involve gauging product tanks. Line managers typically give particular emphasis to startup and shutdown of a particular piece of equipment, mobile equipment, or unit. Managers should consider the following when observing staff to maintain behavioral standards and adherence to operating procedures:

- Changes in technology or process
- Revisions to procedures

- Prioritizing, scheduling, and completing observations regularly
- Involving employees in observation and standards revision
- Staffing levels
- Lessons learned from any injury or incident.

Many organizations have begun to adopt leading indicators that show conditions, behaviors, or activities that show how the safety process is actually working. A prime indicator for this is behavior observations. Besides the obvious advantage of recording "who" was observed, what was observed and the location observed, safe behavior observations also provide you with more advantages as well. These include:

- By using representative sampling, collecting safe and unsafe behavior observations can provide you with a ratio of safe versus unsafe. For example, would you be more concerned about a ratio of 50% unsafe in electrical or 2% unsafe in electrical?
- By actually counting a representative amount of behavior safety observations and not just checking a box for the entire project, you can determine the context of your findings as well. For example, you find three unsafe observations for failure to use safety glasses. Now, if there were only three workers observed, then this is significant. If there were 300 other workers who were wearing safety glasses, then the gravity of the findings is diminished, allowing you to focus on more severe findings.
- Safe behavior observations should be conducted in such a way that allows positive feedback to be employed. The idea is for the supervisor to serve as coach to improve and avoid the stereotypical safety "cop" scenario and creating a "gotcha" mentality of busting workers for safety violations—that only creates an "us versus them" environment and it is not very conducive to teamwork or improving performance. Done in a sincere manner where it is clear that the supervisor is attempting to show genuine care and concern for the safety and well-being of the individual, the line manager's efforts in effective behavior modification are more likely to be met with success. Commending and correcting as the situation warrants needs to be done in a very personal manner such that it is well received by the worker being observed and acted upon in a positive upbeat manner without drama or hard feelings. This should be relatively straightforward if only because safety is a very personal matter and each individual with a survival instinct will recognize that doing things safely is in his or her own best interests.
- Only through safety observations can you measure improvement. Let's say you found a lot of unsafe observations for a certain hazard and implemented an action plan to address it. How would you know it got better? *Keep in mind that an absence of unsafe observations could mean nobody is looking!* An improved ratio of safe versus unsafe should support an improvement in the targeted, specific safety process you were concerned about.

A form typically used in the process industry is shown in Figure 2.1.

Incident Investigation

Some type of incident investigation is a part of nearly every safety management system. Yet, the purpose for doing investigations is often poorly understood. As a result, they can degenerate into finger pointing, assigning blame, and fault finding exercises which seldom determine the real reasons for what happened or arrive at any effective solutions to process of identifying the underlying or basic causes. Even when the purpose is properly defined, decisive, insightful and logically supported analysis can vary greatly in quality from investigation to investigation.

Behavior Based Safety Observation Form				
□ BBS Observation	□ Unsafe Act	□Unsafe Condition	□ Environment	□ Recognition

OBSERVATION:

Employee's Action Taken or Recommendation:

Supervisor or Management Action Taken:

Safety Observation Critical Factors: S=Safe, U=Unsafe/Concern

PPE/Procedures/Methods			Body Position/Mechanics			Slips/Trips			Work Environment		
S	U	Feature	S	U	Feature	S	U	Feature	S	U	Feature
□	□	Eye and Head	□	□	Proper Position	□	□	Proper Footwear	□	□	MSDS, if needed
□	□	Hand and Body	□	□	Assistance Used	□	□	Hazard Awareness	□	□	Lockout
□	□	Footwear	□	□	Use Dolly	□	□	Prompt Cleanup	□	□	Tools Adequate
□	□	Trained on Task	□	□	Smaller Loads	□	□	Tripping Hazards	□	□	Adjacent Work
□	□	Work Permit/JSA	□	□	Don't Twist Body	□	□	Not Rushing	□	□	Signage, if needed
□	□	Trained in BBS	□	□	Get Close to Item	□	□	Step Conditions	□	□	Spill Control

Observers Feedback Given to Other Employees:

Location:	Observers Name:	Date:

Your concerns for safety and suggestions on how to improve our safety program are important. Use this form to submit either safety improvement input and/or a BBS Safety Observation. Your name is optional and the name of the person being observed is not to be used. This information will be used to continually improve our safety system and conditions.

Figure 2.1 Behavior based safety observation form.

Who Should Investigate?

Which supervisors or other managers should be involved with conducting an investigation? Designating an investigator or investigation team is critical first step. As with any type of problem solving, the person with the most interest in the problem may seem as the likely candidate for selection to lead it. But by the same token, it is also important for the investigator to stay objective. The findings have to be truthful and relevant of the problem may not really be solved. The front line supervisor is often directly involved in leading all minor injury incident investigations (i.e., root causes) and near miss events as well. Sometimes the circumstances of an incident may cross organization boundaries and this may require higher level participation by middle or higher level managers. Involvement of staff is often an option too as there are times where the insights of subject matter experts can be instrumental in providing key insights or be involved in forensic analysis as necessary.

Effective investigations will normally:

- Describe what happened.
- Determine the immediate causal factors.
- Identify the underlying, basic or root causes.
- Provide sound recommendations to prevent recurrence.

Cost of Accidents

The Iceberg model (See Figure 2.2) has been used to portray the relationship between the circumstances that result in incidents that tend to be most visible in an organization and those that seem to lurk just beneath the surface. Because most organizations have an active safety management system that will have line management tuned in to identifying and correcting unsafe acts and conditions, those circumstances that result in the more visible loss incidents such as fatalities, lost time injuries, property loss, equipment damage, lost production, and customer complaints tend to occur with much less frequency to learn from past experiences and therefore avoid repeating past mistakes. Incidents should always be viewed as opportunities to improve management systems rather than the other less visible circumstances such as major opportunities to assign blame. Even near misses, minor near misses or close call situations, or those such as problem areas that might include at risk behaviors, quality control issues, equipment upsets, unsafe acts or conditions, human error, etc. One of the things often highlighted in presenting the Iceberg model is that regardless of the eventual which could have had catastrophic consequences or outcome of the

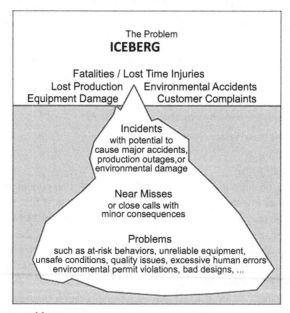

Figure 2.2 Iceberg problem.

situation, they all generally involve the same underlying control factors. Once the sequence of, meaning that the incident did not actually occur, are events has occurred, the type and degree of loss are somewhat a matter of chance. The effect may range from insignificant to catastrophic, from a scratch or dent ranging up to more serious property loss or loss of life. The type and degree of loss depend partly on fortuitous circumstances as well as those efforts to minimize loss. After actions to control and contain the incident have been taken, efforts will usually begin to initiate the incident investigation. Receptivity to the findings from investigation of major incidents have caused process safety management philosophies to evolve, and advance, and have prompted standards that help improve effectiveness of emergency preparedness and incident management as well as aspects of business continuity. Once an incident occurs, it is important to provide prompt administration of first aid and medical care, fast and effective firefighting, prompt shut down, repair and restoration of damaged equipment, the implementation of emergency action plans and rehabilitation of people to return to work.

Organizations must have adequate capability to investigate incidents which occur at their facilities. A capable team should be assembled by the organization which is trained in the techniques of investigation, how to conduct interviews of witnesses, assemble needed documentation, and prepare the investigative report. A multi-disciplinary team is better able to gather the facts of an event, analyze the information, and develop plausible scenarios as to what happened and why. Team members should be selected on the basis of their training, knowledge, and ability to contribute to a team effort to fully investigate the incident. Today it is recognized that operating, design, and maintenance engineers, and managers all have roles in process safety and they should also participate, as appropriate, in incident investigations.

The root cause, of incidents, refers to a variety of problem-solving methods that attempt to identify and correct a problem's root causes. It assumes that the best way to solve problems is by eliminating their root causes. It also works under the belief that addressing obvious symptoms, sometimes referred to as causal or contributing factors, only serves as a short-term solution and does not prevent the problem from happening again.

Root cause analysis (RCA) has been applied across industry sectors and disciplines. Loss prevention and safety engineers have used it to conduct accident analysis; production and maintenance engineers have used it to investigate failures in manufacturing. RCA has also emerged within change management, risk management, systems analysis, and other areas of business.

The process involves defining the problem, investigating through gathering evidence, identifying root causes, implementing solutions and, finally, monitoring those solutions to ensure they continue to prevent the original problem.

At its most basic, the process asks three questions, which together provide the framework of a root cause analysis investigation:

1. What was the problem?
2. How did it happen?
3. Why did it happen? (i.e., causal factors/immediate causes/contributing causes)
4. What were the (multiple) root causes of the problem?
5. What actions should be taken to prevent the problem from occurring again?

Root cause analysis can use a variety of techniques to uncover (multiple) root causes, including cause mapping, change analysis, the Ishikawa fishbone diagram (See Figure 2.3), 5-Whys, and others. These all are designed to analyze the elements affecting a particular outcome to determine the root causes. Root causes mainly focus on problems with materials, machines and equipment, environmental factors, management, methods, and procedures. Root causes are typically related to deficiencies in management systems.

Regardless of the specific technique, all root cause analysis investigations share certain common attributes.

Every cause uncovered by RCA must be backed up by factual evidence and these conclusions derive from careful analysis of "what happened" and "why".

RCA usually uncovers a system of root causes. There is rarely one, singular root cause. Equipment may experience downtime due to a part failure. But why did the

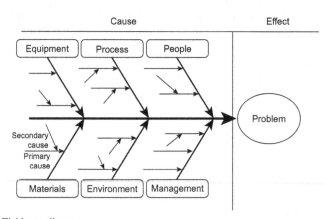

Figure 2.3 Fishbone diagram.

part fail? It could be a combination of part design problems, machine overuse, and improper machine maintenance.

RCA uncovers specific causes and effects. For instance, RCA does not stop with problems such as human error. Instead, it goes beyond this by asking what exact error was committed and why it happened.

RCA results in executable, quantifiable solutions that may be monitored.

RCA does not point blame at any one person or group, but simply identifies causes and effects at the management system level that lead to an incident.

A common finding from major incidents is the failure to adequately communicate lessons learned at one facility to other facilities in the same organization. Sometimes, valuable insights are not captured, shared or applied to the fullest extent possible, resulting in lost opportunities to prevent a repeat incident of a similar nature. Coincidentally, with the participation of personnel from other facilities in the incident investigation, they may point out useful insights or even provide unique insights from incidents at other locations.

Iceberg Model—Safety experts often use the analogy of an iceberg to illustrate the many costs that can arise as a result of a workplace incident. To help conceptualize and understand the relationship between direct and indirect costs, a pioneer in safety named Frank E. Bird, Jr. recognized that injury and illness costs are only a relatively small part of the total cost an incident represents to an organization. He presented a summary of these costs using the diagram of an iceberg to illustrate the fact that the costs associated with injury were only a very small part of the overall costs the so-called "the tip of the iceberg." And while these costs can be very significant destroyers of profit, they are dwarfed in significance by the costs beneath the surface—which he indicated range from at least six to fifty three times as much. His point was that any organization that determined its cost of its accident losses solely in terms of injuries and occupational illnesses (e.g., workers compensation) is looking at only a small fraction of it costs—which come straight out of profit. He keenly observed "Save a dollar in accident costs and you add a dollar straight to profit." The iceberg model has been used to explain the relationship between the direct (calculated) and the indirect costs indicated by the massive part of the iceberg below the waterline—those that are difficult or almost impossible to see (calculate). This diagram has been used extensively and can be accessed on the Occupational Safety and Health Administration (OSHA) website at "Safety Pays." While it's often easy to forget about indirect costs because they're hard to measure, the US OSHA estimates they average between 1 and 4.5 times direct costs!

1931 H.W. Heinrich Accident Study

A now famous accident study by H.W. Heinrich showed that for every accident resulting in a serious injury, there are approximately 29 resulting in only minor injuries and 300 producing no injuries. This discovery led him to conclude that by reacting to only major-injury incidents it is possible that 99.7% of the incidents that occur in an organization are being ignored. Heinrich stressed the fact that the same factors causing a near-accident at one time can cause a major injury the next. By looking at only major injuries in a company's operations management is likely to miss many opportunities to find and

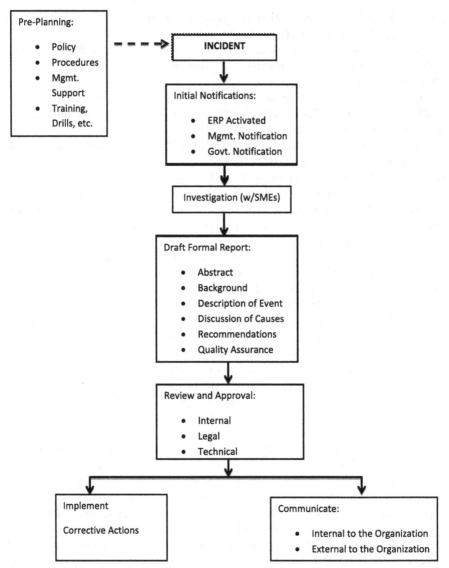

Figure 2.4 Incident investigation flowchart.

eliminate the causes of incidents that nearly result in injury and/or property loss incidents. Effective hazard or loss control requires being aware of the possibilities for all types of incidents and knowing how to prevent them from occurring (see Figure 2.4). Adding the operational discipline slant to this issue should suggest that focus further down the problem Iceberg on things like at risk behaviors (unsafe acts) and unsafe conditions, quality problems, human errors, and issues involving equipment failure can help improve operational discipline and work toward improving performance from the bottom up.

The worldwide petroleum and chemical insurance market estimates for the period 1993 to 2013 (approx. last 20 years) there have been about 1100 major insurance claims (i.e., major incidents) amounting to approximately $32 billion (for property damage and business interruption). Their analysis estimates that the worldwide risk has been constant over this period, i.e., the average frequency and cost impact has been a constant trend, neither increasing or decreasing. This equates on average to 110 losses totaling $2–3 billion per year. Additionally these losses would fit a traditional loss incident ratio triangle (see Figure 2.5) with ever increasing number of losses as the magnitudes of the losses decreases (i.e., as the steps in the triangle widen).

Currently new projects are now in the region of greater than $50+ billion, which equates to the Deepwater Horizon incident loss (April 20, 2010), and the potential for even larger losses from a single incident is still a possibility. The industry must do more to prevent these incidents and improve so that these loss trends decrease.

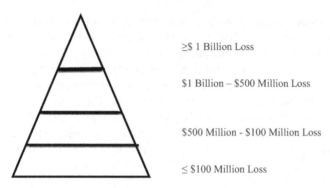

≥$ 1 Billion Loss

$1 Billion – $500 Million Loss

$500 Million - $100 Million Loss

≤ $100 Million Loss

Figure 2.5 Loss triangle, number versus magnitude.

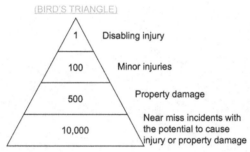

Figure 2.6 Incident ratio triangle.

Accident (Incident) Ratios

Various studies have identified that a tiered relationship exists between the severity of an incident and its frequency of occurrence (Heinrich 1931, Bird 1966, British Safety Council 1975, UK HSE 1993, IET 2009). These studies have concluded that for major injuries (i.e., fatal or serious), a large number of minor injuries and numerous non-injury events typically occur. A definitive numerical relationship has not be established (e.g., 1 to 30 to 300 or 1 to 10 to 600), but the principle and general magnitudes have been accepted by the industry and safety professionals (See Figure 2.6). One of the more important points this drives home is that there are more opportunities to identify incident causes by increasing the focus on incidents at the lower end of the triangle where there are statistically more opportunities to learn and apply corrective actions.

Near Miss Reporting

These are studied and evaluated so that by taking action to prevent near misses, the more serious, moderate, and minor types of incidents can be avoided. Additional reasons for investigating near misses include:

Table 2.1 **Enabling Element Comparisons—A Company's Progression from SMS to OEMS**

Safety Management System Elements	Operational Excellence Management System Elements
Element 1 Leadership and accountability	Element 1 Leadership and accountability
Element 2 Risk assessment and management	Element 2 Customer focus
Element 3 Communications	Element 3 Human resources
Element 4 Competency and training	Element 4 Asset management
Element 5 Asset integrity	Element 5 Process management
Element 6 Safe operations	Element 6 Financial resources
Element 7 Contractors, suppliers and others	Element 7 External services
Element 8 Emergency preparedness	Element 8 Policies and strategies
Element 9 Incident reporting and analysis	Element 9 Information and documentation management
Element 10 Community awareness and off the job safety	Element 10 Change management
Element 11 Continuous improvement	Element 11 Risk management
	Element 12 Learning and continuous improvement
	Element 13 Corporate social responsibility

Safety Management System Enabling Elements

As mentioned earlier, the way you manage safety is the way you manage everything. And because of the inseparable nature of various functions across every organization and the way each supports the other, it follows that they also serve to support the organization's safety performance as well. An organization requires human resources to get the work done to achieve operational goals and objectives; and it's these same resources that must complete their tasks safely and without incident. An organization's engineering resources are involved with establishing standards that affect an organization's assets through the entire asset lifecycle—design, construction, operation, maintenance, and finally decommissioning. Similarly, these same standards are developed in such a manner as to assure safe operation throughout the asset's lifecycle.

Because safety management is integral to operational excellence, it is not surprising that they will also share many of the same enabling elements. Enabling elements in both the safety management system and operational excellence management system can broadly be described as the themes grouping the general organizational, individual, and transactional attributes or aspects necessary to sustain leading performance.

Using the example of leadership and accountability, typically one of the key elements that makes the desired performance across the broader spectrum possible, you will find that they generally drive the line management down a similar path. In both of the examples noted above (see Table 2.1) there were similar provisions mapped out in each element that involved setting the vision, mission, strategies, supporting the overall management system, allocating resources, and extending accountability to assure performance. Essentially the same can be said for the elements titled Risk Management and Continuous Improvement as well. To some degree the title that is assigned to the enabling element is not as important as the way the objectives are articulated and expectations communicated along with the structure, organization, and mapping the supporting processes, procedures, and practices and finally the collective efforts made to bring the system to life and achieve the targeted performance. We will get more into this later.

Interrelationships between SHE (PSM/SMS) and OE Systems

Keywords

Attributes; Audits; Black belts; Characteristics; Continuous improvement; Deepwater Horizon; Elements; Enabling Elements; Ethics; Feedback; Focus areas; Kaizen; Key performance indicators (KPIs); Leadership; Lean; Processes; Resources; Safety case; Self-assessments; Six Sigma; Strategy; Total quality management (TQM); World Class.

There are many references that influence the structure and organization of commonly applied management system frameworks today. Depending on the industry segment and locale(s) an organization chooses to operate in, the development of an organization's own respective operational excellence management system (OEMS) may be influenced by one or more of them. Some of these will include:

- American Chemistry Council (ACC) Responsible Care® Management
- American Institute for Chemical Engineers (AIChE) Center for Chemical Process Safety (CCPS) Risk Based Process Safety Management (PSM) system
- Control of Major Accident Hazards, UK Health and Safety Executive
- DSEAR—Dangerous Substances and Explosives Atmospheres Regulations, UK Health and Safety Executive, 2002
- ISO 9001: 2008, Quality Management Systems, International Organization for Standardization
- ISO 14001: 2004 Environmental Management Systems, International Organization for Standardization
- Occupational Health and Safety Management Systems, OSHAS 18001 (likely to transition over to ISO 45001 (pending—estimated by October 2016)).
- SEVESO-II, Control of Major Accident Hazards, Involving Dangerous Substances, Council of the European Union, Council Directive 96/82/EC, 09 December 1996, amended November 2008
- Successful Health and Safety Management (HSG65), UK Health and Safety Executive 1997
- UK Food Standards Agency regulations
- UK Offshore Installations (Safety Case) regulations 1992 S11992/2885
- US Department of Energy (DOE) Order 5480.19
- US Environmental Protection Agency (EAP) risk management program (RMP) rule 40 Code of Federal Regulations part 68
- US Food and Drug Administration (FDA) Federal Food, Drug, and Cosmetic Act, Current Good Manufacturing Practices
- US Occupational Safety and Health Administration (OSHA) PSM regulation 29 Code of Federal Regulations part 1910.119

Operational excellence (OE) and PSM are closely related. PSM elements typically coincide with the enabling elements selected for inclusion in the OE framework for organizations operating with the oil, gas, and petrochemical industries. Foremost among them are culture, procedures (all types) training, competency, and management review. One of the most fundamental requirements for an effective OEMS is

Applied Operational Excellence for the Oil, Gas, and Process Industries. http://dx.doi.org/10.1016/B978-0-12-802788-2.00003-8
Copyright © 2015 Elsevier Inc. All rights reserved.

consistent execution of procedures and safe work practices. For this to happen, there must be written procedures to execute and workers must be trained in the proper execution of the procedures and the line management chain of command will work closely with the workforce to ensure they have the necessary resources and fully meet the expectations established by company standards.

The Characteristics of Management Systems

For the sake of clarity we will define the OEMS as the collection of focus areas, enabling elements, expectations, processes, programs, procedures, activities, practices, methods, tools, technologies, and organizational capabilities by which organizations consistently and systematically achieve its goals and objectives. The management system is structured, and designed in such a way that there is a logical organized framework with clear documentation that clearly articulates the purpose, intent and expectations that are in turn used to communicate, coordinate, manage and direct the efforts of the organization in a methodical and systematic manner. Companies that develop their OEMS framework do so for the sake of establishing and maintaining performance standards and typically use a continuous improvement strategy to build upon successful execution. This is done in a way that drives improvement in performance in a sustainable fashion. Management systems put in place to meet OEMS expectations will typically incorporate the following characteristics to be effective. It is important that these characteristics be documented.

Scope and Objectives

Scope defines the System's boundaries and identifies its interfaces with other systems, organizations and facilities. Objectives define the systems purpose and expected results.

Enabling Elements

The OE Focus Areas and Enabling Elements are high level categories that logically segregate many of the day-to-day activities, in a manner to ensure that OE expectations are translated into action and excellence is reflected in everything the organization does. These enabling elements are used to collectively define the OEMS and provide the basis for the entire framework.

Processes, Programs, Procedures, and Activities

A business management process is commonly described as a series of related functions and associated actions that are needed in order to do something or achieve a result. A "process" can be defined as the activities by more than one person to produce an outcome or product. It can also be described as an arrangement of various operations that systematically produce a product or products; characteristics include activity of more than one person, suggesting it is collaborative in nature. Formally documented business processes define and address the actions and functions of management to systematically achieve a desired result. The formal process documentation will succinctly describe the various functions of the line manager to achieve his results. A process is one of several key management tools used in the systematic management of OE and the safety management system.

Management process attributes and characteristics usually include:

- Clearly defined provisions detailing ownership, accountabilities, roles, and responsibilities.
- Clearly defined, documented business purpose, scope, objectives, performance standards, and alignment with organizational business strategy.
- Clearly defined relationships and interfaces with related processes, programs and activities.
- Clearly defined standards, e.g., measured quantities, activities, sequences, workflows, and/or procedures.
- Supporting methods, tools, technologies, and documentation.
- Supporting organizational capabilities (e.g., training, skills, competencies, and capacity).
- Clearly defined provisions, practices, and protocols for process performance measurement, management, and continuous improvement.

According to the Cambridge Business Dictionary the term "program" can be described as "a planned series of related events or activities." Procedures are applied by employees as they address the step by step key tasks required in the work. The Cambridge Business Dictionary defines the term "activity" as "the work of a person, group or organization to achieve something, especially to make money." It also refers to a situation where many things are happening or being done.

Responsible and Accountable Resources

This includes approval authorities, experience and training requirements that qualify people to undertake their roles and responsibilities which are specified for both the implementation and execution of the System.

Verification and Measurement

A system must be checked to see whether it is functioning as designed and is achieving its stated purpose. There are two components. Verification determines that processes and procedures are functioning and being effectively executed. Measurement confirms the quality of a System process and determines that the System's objectives and results are being achieved.

Feedback and Improvement Mechanisms

These mechanisms help ensure that actions are taken continuously to improve the System. They use findings from self-assessments (gap analysis), and from verification and measurement activities (internal audits, inspections, reviews, observations, etc.) to enhance the System.

OE is an attribute of organizational leadership that stresses the application of a variety of principles, multiple inter-organizational management systems, and tools toward the sustainable improvement of all operations. Effectiveness is often tracked through the use of key performance indices (KPIs). The process involves focusing on customer needs, keeping employees positive and empowered and continuous improvement of activities in the workplace.

Why is OE centered around health, safety, and environmental management systems? In the oil, gas, and petrochemical industry it is said that how you manage safety is how you manage everything. And with respect to the critical sensitive nature of getting things right in this core area of managing the business it is almost second nature to process industry professionals. When dealing with materials that are to some degree hazardous by their nature (e.g., flammable and combustible nature of hydrocarbons in their various forms), it is imperative to manage these areas of the business properly if for no other reason than the cost involved with getting it wrong can be so astronomically high in both monetary and human terms. In the oil, gas, and petrochemical industry, safety incidents can escalate into environmental disasters and in some cases may also adversely impact human health. A high profile and widely communicated incident that serves as an example of the first can be seen in the earlier referenced BP Deepwater Horizon incident that occurred offshore on its Macondo prospect. That incident, originating as a safety related incident (loss of well control) quickly escalated into one of the largest oil spill disasters (environmental disaster) that the industry has ever seen. Similarly, as an example of the latter, is the previously discussed early morning incident in Bhopal India on December 3rd, 1984. In that incident, involving a loss of containment at the Union Carbide facility resulted in, a release of methyl isocyanate (MIC) gas. The release from the plant exposed more than 500,000 people to MIC and other chemicals. The death toll and level of serious injury that resulted from this incident make it one of the world's worst industrial disasters in history. Union Carbide was sued by the Government of India and agreed to an out-of-court settlement of US$470 million in 1989. A couple key consequences of the Bhopal incident were both a badly tarnished reputation and weakened market position. This ultimately led to a number of divestments and ultimately the remaining Union Carbide assets were acquired by Dow Chemical.

Structure and organization are absolutely key. The management consulting indus-try has established a prosperous niche for itself over the years by coaching and advis-ing management teams in numerous organizations on the benefits of both strategic and tactical efforts to improve the effectiveness of management processes. Some of those methodologies that have achieved fairly wide recognition include the following:

- Lean—Lean is a customer-centric methodology used to continuously improve any process through the elimination of waste in everything you do; it is based on the ideas of "Con-tinuous Incremental Improvement" and "Respect for People." To create a sustaining Lean organization, you lead differently. Lean leaders know the only way to truly understand what is happening is to go to the place where the action occurs.
- Six Sigma—Six Sigma is a philosophy and methodology for managing processes and per-formance. It goes beyond just isolated improvement projects, and is best implemented using a system of components designed to clarify an organization's goals, stakeholders, and needs. Six Sigma is most successful when leadership is truly committed to the philosophy and methodology it entails. In larger companies, a Director or other high-level employee takes the lead role in creating and guiding Six Sigma efforts. Also for large-scale efforts, Black Belts should be trained up front as they will be responsible for leading improvement proj-ects, and in some cases for advising process owners on establishing appropriate metrics and procedures.
- Total quality management (TQM)—TQM consists of organization-wide efforts to install and make permanent a climate in which an organization continuously improves its ability to deliver high-quality products and services to customers. While there is no widely agreed-upon approach, TQM efforts typically draw heavily on the previously developed tools and techniques of quality control. TQM enjoyed widespread attention during the late 1980s and early 1990s before being overshadowed by ISO 9000, Lean manufacturing, and Six Sigma.
- Kaizen—When Japan was devastated after World War II, so was most of its industry. Japan needed a plan, a way to dig itself up from the rubble, to become a world industrial leader. The plan the Japanese used to successfully do so was called Kaizen. Kaizen is an optimistic view that allows people, no matter their station at work, to band as a team to overcome obsta-cles. This is done through constantly fixing what happens. Under the Kaizen model, prob-lems are opportunities, not something to be covered up or explained away. Solving problems strengthens the company over time; covering up problems or spinning them merely puts on a temporary Band-Aid. Under the Kaizen model, problems are opportunities, not something to be covered up or explained away. Solving problems strengthens the company over time; covering up problems or spinning them merely puts on a temporary Band-Aid. Under Kai-zen, if there are no problems, that's a problem; if you don't recognize what your problems are, you won't be able to solve them Eliminating what doesn't work and even improving on what does. Quality control is a big part of the Kaizen concept. Managers who use Kaizen concern themselves first with the way things are done before they focus on results. Western management tends to think of results first, the Kaizen concept dictates that you can't expect to get good results until you improve your process for achieving your desired outcome. For example, in a process-oriented model, workers are rewarded through recognition for following a good process that consists of discipline, time management, skill development, participation, morale, and communication.

To a certain extent these all represent variations on a common theme and involve looking closely at many of the essential management functions at the heart of a "process" as we referred to earlier. They generally include those functions that line management must master to establish and maintain standards necessary to safely and effectively manage their respective operations. As a reminder, key process functions often encompass—*responsibilities, standards, documentation, training, measurement and effectively extending accountability.*

There is also a price/cost component in others that often skews emphasis on "cheapest/fastest" as opposed to cost-effective quality solution. OE models will typically focus on proper design, construction, maintenance, and operation with a focus on establishing and maintaining operational discipline. One of the basic challenges in achieving operational discipline is that of effectively managing people and ensuring they routinely and reliably apply the proper decisions and behaviors at the right time to produce the desired results. This requires a laser-like focus by line management to assure the organization is systematically, methodically, uniformly, and correctly applying company standards for the sake of doing things the right way; each task the right way—every time.

Most major process industries have moved or begun moving towards the OE concept. Accordingly, safety and quality management systems have had to evolve and integrate with the introduction of OE. Not surprisingly most safety management systems already possessed many of the OE features. The key challenge in integrating or adapting to the OE from safety management (as some of the element of each overlap) is not to lose focus of the importance of loss prevention as a key objective or goal of the organization. Also OE recognizes that leadership is single largest factor for its success within an organization. These leaders establish the overall vision and set objectives that challenge the organization to achieve World-Class results.

Benefits of Integrated PSM/SMS/ SHE and OE Programs

Keywords

Asset integrity management system (AIMS); Audits; Compliance; Contractors; Cost saving; Culture; Emergency preparedness; Enterprise risk management (ERM) system; Environmental management system (EMS); Increase efficiencies; Inspection; Investigation; Managing change; Mechanical integrity; Occupational Safety and Health Administration (OSHA); "One-Step Merger"; Operating procedures; Pre-startup safety review (PSSR); Process hazard analysis (PHA); Process safety management (PSM); Regulatory; Responsibilities; Risk Management Plan (RMP); Roles; Training.

While most organizations that adopt the OEMS framework do so by establishing the framework at the corporate level, the way it is applied and administered across the company is of equal if not greater importance. After all, the effectiveness of the system and the value it brings to the enterprise is impacted by the manner in which it is applied in practice at the local level. If it is only dead paper on a shelf somewhere, it may be of questionable value. When it serves as a living document that is embraced and fully supported up and down the chain of command, it has the ability to bring a balanced approach to operating and provide consistency, uniformity, and operational discipline to the individual operations across an organization. Successful implementation of a local OEMS will help to deliver:

- Clear roles and responsibilities for the workforce.
- Clear accountabilities with unambiguous goals.
- A competent workforce that understands and applies entity values and correctly establishes priorities.
- Entities with the resources and capability to implement the entity plan and systematically improve operating processes and activities.
- Compliance with applicable legal and regulatory requirements, and conformance to established company standards and best practices.
- Effective personal and process safety management.
- Continuous risk reduction and disciplined behavior at all levels as well as a willingness to identify potential hazards and accept challenge by intervening to address and properly mitigate unsafe acts and conditions.
- Leadership that listens and responds openly to the workforce and stakeholders.
- Continuous improvement, learning from ourselves and others to improve the leadership, capability and capacity of the organization.
- Customer and stakeholder expectations being met or exceeded.

Energy majors with operations across multiple oil/gas and petro-chemical industry sectors strive to manage in such a way that:

- Intensifies their focus on protection and preservation of the workplace and the community health, safety, and the environment.

Applied Operational Excellence for the Oil, Gas, and Process Industries. http://dx.doi.org/10.1016/B978-0-12-802788-2.00004-X
Copyright © 2015 Elsevier Inc. All rights reserved.

- Further solidifies their competitive positions as reliable suppliers of products and commodities to world markets.
- Enhances and sustains asset and process reliability, efficiency, cost effectiveness and profitability in meeting social and commercial commitments to internal and external customers.

Within the major industries, Operational Excellence refers to functional efforts required to assure the reliable, efficient, cost-effective, and profitable development, production, and delivery of products and services in ways that consistently meet or exceed the expectations of the stakeholders and customers in a safe, healthy, and environmentally and socially responsible manner. OE is achieved through coordinated efforts, excellent communication, and a lot of hard work. Beyond that it involves sustaining best in class results and establishing and maintaining as well as constantly improving the excellent practices in areas such as leadership, people management, policies and strategies, processes, resources management, and customer focus.

Over the years, a number of leading companies have developed individual management systems to address the various challenges that are often pulled together under an OE umbrella for the sake of comprehensively addressing today's business challenges. While over time with the way the individual disciplines tended to be more organizationally segregated (e.g., safety staff in a separate organization from environmental services, health and occupational medicine, engineering services, maintenance human resources, internal audit, etc.) they may have given the appearance of being somewhat compartmentalized (with something of a silo effect) isolating one from the other. Getting everyone together on the same page was often challenging for senior management, proponents often grew audit weary from multiple audits/reviews and the effort usually involved considerable coordination, communication and cooperation to maintain a uniform path forward. Over time many companies have gone through reorganization efforts to tear down the walls that may have existed between internal staff organizations and have strategically combined staff organizations such as Safety, Health and Environment so as to streamline intercompany communications and strategic direction and better support operating organizations and the line management chain of command. Structuring the OE program in a comprehensive integrated fashion helps unite these efforts with a common purpose and gain additional synergy and efficiency. For example, some companies have long established safety management systems that address and incorporate mechanical integrity as part of the process safety management efforts. But mechanical integrity to assure safety can go beyond the basic efforts to assure plant and equipment remain fit for service, asset integrity management systems (AIMS) incorporate aspects of inspection, preventive maintenance, as well as other factors relating to cost effectively managing the assets throughout the asset life cycle that can also impact cost and schedule.

Various other management systems may exist within different silos in an organization with little if any interconnection, to the point where they are all treated as unique and distinctly separately managed issues and concerns, in spite of the fact that performance in one area can have a significant impact in performance within another. In many instances there are an equal number of auditing programs going simultaneously that leave individual facility management teams audit weary. Examples may include a company's safety management system (SMS), the environmental management system

(EMS), the enterprise risk management (ERM) system, the AIMS, and others and these can be integrated to provide important support aimed at specific OE objectives. An OE Implementation Plan typically includes: short and long-term strategies; and plans to better integrate, align and eliminate duplication; add essential processes/practices; and consolidate the design, implementation, and operation of these existing systems to support all OE objectives.

Integrating Process Safety Management–The Purpose Behind PSM

The major objective of process safety management (PSM) of highly hazardous chemicals is to prevent unwanted releases of those chemicals especially into locations that could expose employees and others to serious hazards. An effective process safety management program requires a systematic approach to evaluating the whole chemical process. Using this approach, the process design, process technology, process changes, operational and maintenance activities and procedures, non-routine activities and procedures, emergency preparedness plans and procedures, training programs, as well as other elements that affect the process are incorporated in the evaluation. Functionally integrating PSM into an organization's OEMS requires some logical thought be given not just for where the standards are mapped within the OEMS within each respective element, but how they interrelate and work together in the organization to deliver the desired performance. Placement and structural considerations are just the start, there also needs to be provisions to assure that each requirement is sufficiently detailed, known to the management and workforce, and diligently implemented as well as continuously improved.

Making the Process Industry Safer Through PSM

The various lines of defense that have traditionally been incorporated into the design and operation of the process to prevent or mitigate the release of hazardous chemicals needed to be evaluated and strengthened to ensure their effectiveness at each level. Process safety management is the proactive identification, evaluation, and mitigation or prevention of chemical releases that could occur as a result of failures in processes, procedures, or equipment. The process safety management standard targets highly hazardous chemicals that have the potential to cause a catastrophic incident. In the United States, the Department of Labor's Occupational Safety and Health Administration (OSHA) promulgated the PSM regulation and drafted it in the form of a performance standard. A performance standard differs from a specification standard (e.g., *29CFR1910.24 (d)* "Stair width." Fixed stairways shall have a minimum width of 22 inches.) in that it essentially outlines the *performance* expected of an employer organization to safely achieve the stated objectives. OSHA's purpose for PSM was to meet the objectives laid out under the Clean Air Act of 1990 and to map out the expectations in a framework that could be readily adapted and incorporated into those employers already typically used to manage their operations. Just as the OSHA PSM standard is required by the Clean Air Act Amendments, so too is the Environmental Protection Agency's **Risk Management Plan** (RMP), which was promulgated later. Employer organizations that merge the two sets of requirements PSM and RMP into their organization's OE management system find they can better achieve compliance with each as well as enhance their relationship with the local community.

The overall objective for employer organizations as they implemented the provisions of the standard within the scope of their operations is to prevent or mitigate episodic chemical releases that could lead to a catastrophe in the workplace and possibly impact the surrounding community. To control these types of hazards, many organizations needed to develop the necessary expertise, experience, judgment, and initiative within their work force to properly implement and maintain an effective process safety management performance. For those organizations operating in the oil, gas, and petrochemical industries, the standard served to help formalize many of the best practices that already existed and were in place to one degree or another in many of the leading organizations.

Although OSHA believes process safety management will have (and continue to have) a positive effect on the safety of employees, it will offer other potential benefits to employer organizations, such as increased productivity. While not distinguishing too much between organizational size and structure, those smaller sized businesses with more limited resources might consider alternative avenues of decreasing the risks associated with highly hazardous chemicals at their workplaces, and because a performance-based standard is not a one size fits all solution, it does provide some latitude for companies to develop the solutions that work best for them—providing of course that it delivers the proper performance within the context of the basic framework.

One method that might be considered is reducing inventory of the highly hazardous chemical. This reduction in inventory will result in reducing the risk or potential for a catastrophic incident. Also, employer organizations, (including small employers who are often more nimble because of their size) may establish more efficient inventory control by reducing the quantities of highly hazardous chemicals on site to something below the established threshold. This reduction can be accomplished by ordering smaller shipments and maintaining the minimum inventory necessary for efficient and safe operation. When reduced inventory is not feasible, an organization might consider dispersing inventory to several locations on site. Dispersing storage into locations so that a release in one location will not cause a release in another location is also a practical way to reduce the risk or potential for catastrophic incidents.

Exceptions

The PSM standard will not typically apply in the following situations:

- Normally unoccupied remote facilities,
- Hydrocarbon fuels used solely for workplace consumption as a fuel (e.g., propane used for comfort heating, gasoline for vehicle refueling),

- Flammable liquid stored in atmospheric tanks or transferred, which are kept below their normal boiling point without benefit of chilling or refrigerating and are not connected to a process,
- Retail facilities,
- Oil or gas well drilling or servicing operations.

Process Safety Information Hazards of the Chemicals Used in the Process

Complete and accurate written information concerning process chemicals, process technology, and process equipment is essential to an effective process safety management program and to a process hazard analysis (PHA). The compiled information will be a necessary resource to a variety of users including the team performing the PHA as required by PSM; those developing the training programs and the operating procedures; contractors whose employees will be working with the process; those conducting the pre-startup reviews; as well as local emergency preparedness planners, and insurance and enforcement officials.

The information to be compiled about the chemicals, including process intermediates, needs to be comprehensive enough for an accurate assessment of the fire and explosion characteristics, reactivity hazards, the safety and health hazards to workers, and the corrosion and erosion effects on the process equipment and monitoring tools. Current material safety data sheet (MSDS) information can be used to help meet this but must be supplemented with process chemistry information, including runaway reaction and over-pressure hazards, as applicable.

Technology of the Process

Process technology information will be a part of the process safety information package and typically includes company established criteria for maximum inventory levels for process chemicals; limits beyond which would be considered upset conditions; and a qualitative estimate of the consequences or results of deviation that could occur if operating beyond the established process limits. Companies are encouraged to use diagrams that will help users understand the process.

A block flow diagram is used to show the major process equipment and interconnecting process flow lines and flow rates, stream composition, temperatures, and pressures when necessary for clarity. The block flow diagram is a simplified diagram.

Process flow diagrams (see Figure 4.1 for an example) are more complex and will show all main flow streams including valves to enhance the understanding of the process, as well as pressures and temperatures on all feed and product lines within all major vessels and in and out of headers and heat exchangers, and points of pressure and temperature control. Also, information on construction materials, pump capacities, and pressure heads, compressor horsepower, and vessel design pressures and temperatures are shown when necessary for clarity.

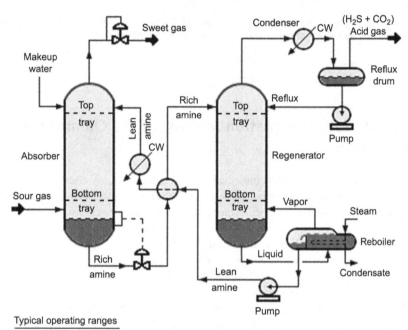

Typical operating ranges

Absorber: 35 to 50 °C and 5 to 205 atm of absolute pressure
Regenerator: 115 to 126 °C and 1.4 to 1.7 atm of absolute pressure
                            at tower bottom

Figure 4.1 Process flow diagram example.
Licensed under CC BY-SA 3.0 via Wikimedia Commons – https://commons.wikimedia.org/
wiki/File:AmineTreating.png#/media/File:AmineTreating.png.

Equipment in the Process

Piping and instrument diagrams (P&IDs) may be the more appropriate type of diagram to show some of the above details as well as display the information for the piping designer and engineering staff. P&IDs are to be used to describe the relationships between equipment and instrumentation as well as other relevant information that enhances clarity. Computer software programs that produce P&IDs or other diagrams useful to the information package are often used to help meet this requirement.

The information pertaining to process equipment design must be documented. In other words, documentation highlights the codes and standards that were relied on to establish good engineering practice and the basis for the design. These codes and standards are published by such organizations as the American Society of Mechanical Engineers, the American Petroleum Institute (API), American National Standards Institute (ANSI), National Fire Protection Association (NFPA), American Society for Testing and Materials (ASTM), The National Board of Boiler and Pressure Vessel Inspectors, National Association of Corrosion Engineers (NACE), American Society of Exchange Manufacturers Association, and Model Building Code groups.

For existing equipment designed and constructed many years ago in accordance with the codes and standards available at that time and no longer in general use today, the organization must document which codes and standards were used and show that the design and construction along with the testing, inspection, and operation are still suitable for the intended use. Where the process technology requires a design that departs from the applicable codes and standards, the organization must document that the design and construction are suitable for the intended purpose.

Employee Involvement

Section 304 of the Clean Air Act Amendments states that employers are to consult with their employees and their representatives regarding their efforts in developing and implementing the process safety management program elements and hazard assessments. It also requires employers to train and educate their employees and to inform affected employees of the findings from incident investigations required by the process safety management program. Many organizations have established methods under their existing Operational Excellence (OE) management system and/ or Safety Health and Environmental (SHE) programs to keep employees and their representatives informed about relevant safety and health issues. Those that don't yet should be able to readily to adapt these practices and procedures to meet their obligations under PSM.

Examples of mechanisms used to trigger employee involvement include establishing safety committees attended by employees and management representatives as well as unit safety meetings, training, involvement in incident investigation teams and others.

Process Hazard Analysis

A PHA, or evaluation, is one of the most important elements of the process safety management program. A PHA is an organized and systematic effort to identify and analyze the significance of potential hazards associated with the processing or handling of highly hazardous chemicals. A PHA provides information that will assist companies and employees in making decisions for improving safety and reducing the consequences of unwanted or unplanned releases of hazardous chemicals.

A PHA analyzes potential causes and consequences of fires, explosions, releases of toxic or flammable chemicals, and major spills of hazardous chemicals. The PHA focuses on equipment, instrumentation, utilities, human actions (routine and non-routine), and external factors that might affect the process.

The selection of a PHA methodology or technique will be influenced by many factors including how much is known about the process. Is it a process that has been operated for a long period of time with little or no innovation and extensive experience has been generated with its use? Or, is it a new process or one that has been changed frequently by the inclusion of innovation features? Also, the size and complexity of the process will influence the decision as to the appropriate PHA methodology to use. All PHA methodologies are subject to certain limitations. For example, the checklist methodology works well when the process is very stable and no changes are made, but it is not as effective when the process has undergone extensive change. The checklist may miss the most recent changes and consequently they would not be evaluated. Another limitation to be considered concerns the assumptions made by the team or analyst. The PHA is dependent on good judgment and the assumptions made during the study need to be documented and understood by the team and reviewer and kept for a future PHA.

The team conducting the PHA needs to understand the methodology that is going to be used. A PHA team can vary in size from two to a number of people with varied operational and technical backgrounds. Some team members may be part of the team for only a limited time. The team leader needs to be fully knowledgeable in the proper implementation of the PHA methodology to be used and should be impartial in the evaluation. The other full or part-time team members need to provide the team with expertise in areas such as process technology, process design, operating procedures and practices, alarms, emergency procedures, instrumentation, maintenance procedures, both routine and nonroutine tasks, including how the tasks are authorized; procurement of parts and supplies; safety and health; and any other relevant subjects. At least one team member must be familiar with the process.

The ideal team will have an intimate knowledge of the standards, codes, specifications, and regulations applicable to the process being studied. The selected team members need to be compatible and the team leader needs to be able to manage the team and the PHA study. The team needs to be able to work together while benefiting from the expertise of others on the team or outside it to resolve issues and to forge a consensus on the findings of the study and recommendations. The application of a PHA to a process may involve the use of different methodologies for various parts of the process. For example, a process involving a series of unit operations of varying sizes, complexities, and ages may use different methodologies and team members for each operation. Then the conclusions can be integrated into one final study and evaluation. A more specific example is the use of a PHA checklist for a standard boiler or heat exchanger and the use of a Hazard and Operability PHA for the overall process. Also, for batch-type processes like custom batch operations, a generic PHA of a representative batch may be used where there are only small changes of monomer or other ingredient ratio and the chemistry is documented for the full range and ratio of batch ingredients. Another process where the company might consider using a generic type of PHA is a gas plant. Often these plants are relatively easy to relocate and are simply moved from site to site. Accordingly, a generic PHA may be used for such movable plants. Also, when a company has several similar size gas plants and no sour gas (containing hydrogen sulfide [H_2S]) is being processed at the site, a generic PHA is feasible as long as the variations of the individual sites are accounted for in the PHA.

Finally, when a company has a large continuous process with several control rooms for different portions of the process, such as for a distillation tower and a blending operation, the organization may wish to do each segment separately and then integrate the final results.

Small businesses covered by this rule will often have processes that have less storage volume and less capacity and may be less complicated than processes at a large facility.

Accordingly, it would not be a surprise to see the application of less complex method-ologies being used to meet the PHA criteria in the standard. These PHAs can be done in less time and with fewer people being involved. A less complex process generally means that less data, P&IDs, and process information are needed to perform a PHA.

Many small businesses have processes that are not unique, such as refrigerated warehouses, cold storage lockers, or water treatment facilities. Where associations have a number of members with such facilities, a generic PHA, evolved from a check-list or what-if questions, could be developed and effectively used by companies to reflect their particular process; this would simplify compliance for them.

When the company has a number of processes that require a PHA, the company must set up a priority system to determine which PHAs to conduct first. A prelim-inary hazard analysis may be useful in setting priorities for the processes that the company has determined are subject to coverage by the process safety management standard. Consideration should be given first to those processes that could potentially have an adverse effect on the largest number of employees. This priority setting also should consider the potential severity of a chemical release, the number of potentially affected employees, the operating history of the process, such as the frequency of chemical releases, the age of the process, and any other relevant factors. Together, these factors would suggest a ranking order using either a weighting factor system or a systematic ranking method. The use of a preliminary hazard analysis will assist the company in determining which process should be of the highest priority for hazard analysis resulting in the greatest improvement in safety at the facility occurring first.

Operating Procedures

Operating procedures describe tasks to be performed, data to be recorded, operating conditions to be maintained, samples to be collected, and safety and health precau-tions to be taken. The procedures need to be technically accurate, understandable to employees, and revised periodically to ensure that they reflect current operations. The process safety information package helps to ensure that the operating procedures and practices are consistent with the known hazards of the chemicals in the process and that the operating parameters are correct. Operating procedures should be reviewed by engineering staff and operating personnel to ensure their accuracy and that they provide practical instructions on how to actually carry out job duties safely. Also the company must certify annually that the operating procedures are current and accurate.

Operating procedures provide specific instructions or details on what steps are to be taken or followed when carrying out the stated procedures. The specific instructions should include the applicable safety precautions and appropriate information on safety implications. For example, the operating procedures addressing operating parame-ters will contain operating instructions about pressure limits, temperature ranges, flow rates, what to do when an upset condition occurs, what alarms and instruments are pertinent if an upset condition occurs, and other subjects. Another example of using operating instructions to properly implement operating procedures is in starting up or shutting down the process. In these cases, different parameters will be required from those of normal operation. These operating instructions need to clearly indicate the distinctions between startup and normal operations, such as the appropriate allowances for heating up a unit to reach the normal operating parameters. Also, the operating instructions need to describe the proper method for increasing the temperature of the unit until the normal operating temperatures are reached.

Computerized process control systems add complexity to operating instructions. These operating instructions need to describe the logic of the software as well as the relationship between the equipment and the control system; otherwise, it may not be apparent to the operator.

Operating procedures and instructions are important for training operating personnel. The operating procedures are often viewed as the standard operating practices (SOPs) for operations. Control room personnel and operating staff, in general, need to have a full understanding of operating procedures. If workers are not fluent in English, then procedures and instructions need to be prepared in a second language understood by the workers. In addition, operating procedures need to be changed when there is a change in the process. The consequences of operating procedure changes need to be fully evaluated and the information conveyed to the personnel. For example, mechanical changes to the process made by the maintenance department (like changing a valve from steel to brass or other subtle changes) need to be evaluated to determine whether operating procedures and practices also need to be changed. All management of change actions are to be coordinated and integrated with current operating procedures, and operating personnel must be alerted to the changes in procedures before the change is made. When the process is shut down to make a change, the operating procedures must then be updated before re-starting the process.

Training must include instruction on how to handle upset conditions as well as what operating personnel are to do in emergencies such as pump seal failures or pipeline ruptures. Communication among operating personnel and workers within the process area performing non-routine tasks also must be maintained. The hazards of the tasks are to be conveyed to operating personnel in accordance with established procedures and to those performing the actual tasks. When the work is completed, operating personnel should be informed to provide closure on the job.

Employee Training

All employees, including maintenance and contractor employees involved with highly hazardous chemicals, need to fully understand the safety and health hazards of the chemicals and processes they work with so they can protect themselves, their fellow employees, and the citizens of nearby communities. Training conducted in compliance with OSHA's Hazard Communication standard (*29CFR1910.1200*) will inform employees about the chemicals they work with and familiarize them with reading and understanding MSDSs. However, additional training in subjects such as operating procedures and safe work practices, emergency evacuation and response, safety procedures, routine and non-routine work authorization activities, and other areas pertinent to process safety and health need to be covered by the company's training program.

In establishing their training programs, companies must clearly identify the employees to be trained, the subjects to be covered, and the goals and objectives they wish to achieve. The learning goals or objectives should be written in clear measurable terms before the training begins. These goals and objectives need to be tailored to each of the specific training modules or segments. Companies should describe the important actions and conditions under which the employee will demonstrate competence or knowledge as well as what is acceptable performance.

Hands-on training, where employees actually apply lessons learned in simulated or real situations, are often incorporated as this is known to enhance learning. For example, operating personnel, who will work in a control room or at control panels, benefit by being trained at a simulated control panel. Upset conditions of various types could be displayed on the simulator, and then the employee could go through the proper operating procedures to bring the simulator panel back to the normal operating parameters. A training environment could be created to help the trainee feel the full reality of the situation but under controlled conditions. This type of realistic training can be very effective in teaching employees correct procedures while allowing them also to see the consequences of what might happen if they do not follow established operating procedures. Other training techniques using videos or training also can be very effective for teaching other job tasks, duties, or imparting other important information. An effective training program will allow employees to fully participate in the training process and to practice their skills or knowledge.

Companies need to periodically evaluate their training programs to see if the necessary skills, knowledge, and routines are being properly understood and implemented by their trained employees. Training is no substitute for supervision, so many company OE or Safety Management systems will incorporate provisions for supervisor procedural observations.

Methods for evaluating the training should be developed along with the training program goals and objectives. Training program evaluation will help companies to determine the amount of training their employees understood, whether the desired results were obtained, and whether they are equipped to properly apply the training. If, after the evaluation, it appears that the trained employees are not at the level of knowledge and skill that was expected, the company would typically revise the training program, provide retraining, or provide more frequent refresher training sessions and instruction until the deficiency is resolved. Those who conducted the training and those who received the training also should be consulted as to how best to improve the training process. If there is a language barrier, the language known to the trainees should be used to reinforce the training messages and information. Some of the means for evaluating such issues and addressing action plans to adequately resolve them are often involved in systematic review and improvement as part of the organization's Continuous Improvement efforts.

Careful consideration must be given to ensure that employees, including maintenance and contract employees, receive current and updated training. For example, if changes are made to a process, affected employees must be trained in the changes and understand the effects of the changes on their job tasks. Additionally, as already discussed, the evaluation of the employee's absorption of training will certainly determine the need for further training.

Contractors

Companies who use contractors to perform work in and around processes that involve highly hazardous chemicals have to establish a screening process so that they hire and use only contractors who accomplish the desired job tasks without compromising the safety and health of any employees at a facility. For contractors whose safety performance on the job is not known to the hiring company, the company must obtain information on injury and illness rates and experience and should obtain contractor references. In addition, the company must ensure that the contractor has the appropriate job skills, knowledge, and certifications (e.g., for pressure vessel welders). Contractor work methods and experience should be evaluated. For example, does the contractor conducting demolition work swing loads over operating processes or does the contractor avoid such hazards?

Maintaining a site injury and illness log for contractors is another method companies must use to track and maintain current knowledge of activities involving contract employees working on or adjacent to processes covered by PSM. Injury and illness logs of both the company's employees and contract employees allow the company to have full knowledge of process injury and illness experience. This log contains information useful to those auditing process safety management compliance and those involved in incident investigations.

Contract employees must perform their work safely. Considering that contractors often perform very specialized and potentially hazardous tasks, such as confined space entry activities and nonroutine repair activities, their work must be controlled while they are on or near a process covered by PSM. A permit system or work authorization system for these activities is helpful for all affected companies. The use of a work authorization system keeps a company informed of contract employee activities. Thus, the company has better coordination and more management control over the work being performed in the process area. A well-run and well-maintained process, where employee safety is fully recognized, benefits all of those who work in the facility whether they are employees of the company or the contractor.

Pre-Startup Safety Review (PSSR)

For new processes, the company will find a PHA helpful in improving the design and construction of the process from a reliability and quality point of view. The safe operation of the new process is enhanced by making use of the PHA recommendations before final installations are completed. P&IDs should be completed, the operating procedures put in place, and the operating staff trained to run the process, before startup. The initial startup procedures and normal operating procedures must be fully evaluated as part of the pre-startup review to ensure a safe transfer into the normal operating mode.

For existing processes that have been shut down for turnaround or modification, the company must ensure that any changes other than "replacement in kind" made to the process during shutdown go through the management of change procedures. P&IDs

will need to be updated, as necessary, as well as operating procedures and instructions. If the changes made to the process during shutdown are significant and affect the training program, then operating personnel as well as employees engaged in routine and nonroutine work in the process area may need some refresher or additional training. Any incident investigation recommendations, compliance audits, or PHA recommendations need to be reviewed to see what affect they may have on the process before beginning the startup.

Mechanical Integrity of Equipment

Companies must review their maintenance programs and schedules to see if there are areas where "breakdown" is used rather than the more preferable on-going mechanical integrity program. Equipment used to process, store, or handle highly hazardous chemicals has to be designed, constructed, installed, and maintained to minimize the risk of releases of such chemicals. This requires that a mechanical integrity program be in place to ensure the continued integrity of process equipment.

Elements of a mechanical integrity program include identifying and categorizing equipment and instrumentation, inspections and tests and their frequency; maintenance procedures; training of maintenance personnel; criteria for acceptable test results; documentation of test and inspection results; and documentation of manufacturer recommendations for equipment and instrumentation.

Process Defenses

The first line of defense a company has is to operate and maintain the process as designed and to contain the chemicals. This is backed up by the second line of defense which is to control the released chemicals through venting to scrubbers or flares, or to surge or overflow tanks designed to receive such chemicals. This also would include fixed fire protection systems like sprinklers, water spray, or deluge systems, monitor guns, dikes, designed drainage systems, and other systems to control or mitigate hazardous chemicals once an unwanted release occurs.

Written Procedures

The first step of an effective mechanical integrity program is to compile and categorize a list of process equipment and instrumentation to include in the program. This list includes pressure vessels, storage tanks, process piping, relief and vent systems, fire protection system components, emergency shutdown systems and alarms, and interlocks and pumps. For the categorization of instrumentation and the listed equipment, the company should set priorities for which pieces of equipment require closer scrutiny than others.

Inspection and Testing

The mean time to failure of various instrumentation and equipment parts would be known from the manufacturer's data or the company's experience with the parts, which then influence inspection and testing frequency and associated procedures. Also, applicable codes and standards—such as the National Board inspection Code, or those from the ASTM, API, NFPA, American National Standards institute, American Society of Mechanical Engineers, and other groups—provide information to help establish an effective testing and inspection frequency, as well as appropriate methodologies.

The applicable codes and standards provide criteria for external inspections for such items as foundation and supports, anchor bolts, concrete or steel supports, guy wires, nozzles and sprinklers, pipe hangers, grounding connections protective coatings and insulation, and external metal surfaces of piping and vessels. These codes and standards also provide information on methodologies for internal inspection and a frequency formula based on the corrosion rate of the materials of construction. Also, internal and external erosion must be considered along with corrosion effects for piping and valves. Where the corrosion rate is not known, a maximum inspection frequency is recommended (methods of developing the corrosion rate are available in the codes). Internal inspections need to cover items such as the vessel shell, bottom and head; metallic linings; nonmetallic linings; thickness measurements for vessels and piping; inspection for erosion, corrosion, cracking and bulges; internal equipment like trays, baffles, sensors and screens for erosion, corrosion or cracking and other deficiencies. Some of these inspections may be performed by state or local government inspectors under state and local statutes. However, companies must develop procedures to ensure that tests and inspections are conducted properly and that consistency is maintained even where different employees may be involved. Appropriate training must be provided to maintenance personnel to ensure that they understand the preventive maintenance program procedures, safe practices, and the proper use and application of special equipment or unique tools that may be required. Inspection activities should be evaluated as part of an OE process evaluation (see Figure 4.2).

Quality Assurance

A quality assurance system helps to ensure the use of proper materials of construction, the proper fabrication and inspection procedures, and appropriate installation procedures that recognize field installation concerns. The quality assurance program is an essential part of the mechanical integrity program and will help maintain the primary and secondary lines of defense designed into the process to prevent unwanted chemical releases or to control or mitigate a release. "As built" drawings, together with certifications of coded vessels and other equipment and of construction, must be verified and retained in the quality assurance documentation. Equipment installation jobs need to be properly inspected in the field for use of proper materials and procedures and to ensure that qualified craft workers do the job. The use of appropriate gaskets,

Critical Measurements	Key Performance Indicators	Quarter X Performance			Detailed Information to Date, 200X			Month Score	G	Y	R	Parameters for the Measurement
	Monthly Results	Month	Month	Month	Out of	Weight	Adjusted	% Score				
INSPECTION PROCESS												
CLOSURE RATE	No. of CARs closed	10	12	5	30	0.2	0.13	18	X			G > 17; Y>14; R<13,9
QUALITY OF CLOSURE	Total no. of deficiencies logged	12	10	8	30	0.05	0.06	5	X			G > 17; Y>14; R<13,9
QUALITY OF CLOSURE	Total no. of repeat deficiencies logged	3	2	2	30	-0.2	-0.06	-4.67		X		No.of items not repeated G > 17; Y>14; R<13,12
ACCURACY	Total no. of deficiencies noted by audit team/ missed by inspection team	4	3	4	30	-0.2	-0.07	-7.33			X	No.of items not missed G > 17; Y>14; R<13,11

Performance Drivers

	Performance Drivers	Month	Month	Month	Out of	Compl.	Incompl.	% COMPLETE	G	Y	R
General EHS	G1 Access	2	1	1	30	3	2	75%		X	
General EHS	G2 Electrical	0	1	2	30	3	0	100%	X		
General EHS	G3 Chemical/ Environmental	5	1	1	30	3	4	43%			X
General EHS	G4 Waste	0	0	3	30	1	2	33%			X
Facilities	Performance Drivers	Month	Month	Month	Out of	Compl.	Month	% Growth			
Special	Performance Drivers	Month	Month	Month	Month Projects	Compl.	Incompl.	% Complete			

X MONTH SCORE: **11.00**

RED < 60%
YELLOW 61 – 80%
GREE 81–100%

Figure 4.2 Inspection scorecard.

packing, bolts, valves, lubricants, and welding rods needs to be verified in the field. Also, procedures for installing safety devices need to be verified, such as the proper torque on the bolts on rupture disc installations, uniform torque on flange bolts (and tightening sequence), and proper installation of pump seals. If the quality of parts is a problem, it may be appropriate for the company to conduct audits of the equipment supplier's facilities to better ensure proper purchases of required equipment suitable for intended service. Any changes in equipment that may become necessary will need to be reviewed for management of change procedures.

Non-routine Work Authorizations

Non-routine work conducted in process areas must be controlled by the company in a consistent manner. The hazards identified involving the work to be accomplished must be communicated to those doing the work and to those operating personnel whose work could affect the safety of the process. A work authorization notice or permit must follow a procedure that describes the steps the maintenance supervisor, contractor representative, or other person needs to follow to obtain the necessary clearance to start the job. The work authorization procedures must reference and coordinate, as applicable, lockout/tagout procedures, line breaking procedures, confined space entry procedures, and hot work authorizations. This procedure also must provide clear steps to follow once the job is completed to provide closure for those that need to know the job is now completed and that equipment can be returned to normal.

Managing Change

To properly manage changes to process chemicals, technology, equipment, and facilities, one must define what is meant by change. In the process safety management standard, change includes all modifications to equipment, procedures, raw materials, and processing conditions other than "replacement in kind." These changes must be properly managed by identifying and reviewing them prior to implementing them. For example, the operating procedures contain the operating parameters (pressure limits, temperature ranges, flow rates, etc.) and the importance of operating within these limits. While the operator must have the flexibility to maintain safe operation within the established parameters, any operation outside of these parameters requires review and approval by a written management of change procedure. Management of change also covers changes in process technology and changes to equipment and instrumentation. Changes in process technology can result from changes in production rates, raw materials, experimentation, equipment unavailability, new equipment, new product development, change in catalysts, and changes in operating conditions to improve yield or quality. Equipment changes can be in materials of construction, equipment specifications, piping pre-arrangements, experimental equipment, computer program revisions, and alarms and interlocks. Companies must establish means and methods to detect both technical and mechanical changes.

Temporary changes have caused a number of catastrophes over the years, and companies must establish ways to detect both temporary and permanent changes. It is important that a time limit for temporary changes be established and monitored since otherwise, without control, these changes may tend to become permanent. Temporary changes are subject to the management of change provisions. In addition, the management of change procedures are used to ensure that the equipment and procedures are returned to their original or designed conditions at the end of the temporary change. Proper documentation and review of these changes are invaluable in ensuring that safety and health considerations are incorporated into operating procedures and processes. Companies may wish to develop a form or clearance sheet to facilitate the processing of changes through the management of change procedures. A typical change form may include a description and the purpose of the change, the technical basis for the change, safety and health considerations, documentation of changes for the operating procedures, maintenance procedures, inspection and testing, P&IDs, electrical classification, training and communications, pre-startup inspection, duration (if a temporary change), approvals, and authorization. Where the impact of the change is minor and well understood, a check list reviewed by an authorized person, with proper communication to others who are affected, may suffice.

For a more complex or significant design change, however, a hazard evaluation procedure with approvals by operations, maintenance, and safety departments may be appropriate. Changes in documents such as P&IDs, raw materials, operating procedures, mechanical integrity programs, and electrical classifications should be noted so that these revisions can be made permanent when the drawings and procedure manuals are updated. Copies of process changes must be kept in an accessible location to ensure that design changes are available to operating personnel as well as to PHA team members when a PHA is being prepared or being updated.

Incident Investigation

Incident investigation is the process of identifying the underlying causes of incidents and implementing steps to prevent similar events from occurring. The intent of an incident investigation is for companies to learn from past experiences and thus avoid repeating past mistakes. Companies recognize and investigate the types of events that resulted in or could reasonably have resulted in a catastrophic release. These events are sometimes referred to as "near misses," meaning that a serious consequence did not occur, but could have. Companies must develop in-house capability to investigate incidents that occur in their facilities. A team should be assembled by the company and trained in the techniques of investigation including how to conduct interviews of witnesses, assemble needed documentation, and write reports. A multi-disciplinary team is better able to gather the facts of the event and to analyze them and develop plausible scenarios as to what happened, and why. Team members should be selected on the basis of their training, knowledge, and ability to contribute to a team effort to fully investigate the incident.

OSHA's recommended incident notification and investigation forms are provided in Figure 4.3.

Employee's Report of Injury Form

Instructions: Employees shall use this form to report all work related injuries, illnesses, or "near miss" events (which could have caused an injury or illness) – *no matter how minor*. This helps us to identify and correct hazards before they cause serious injuries. This form shall be completed by employees as soon as possible and given to a supervisor for further action.

I am reporting a work related: □ Injury □ Illness □ Near miss

Your Name:_____

Job title:_____

Supervisor:_____

Have you told your supervisor about this injury/near miss? □ Yes □ No

Date of injury/near miss: _____Time of injury/near miss:_____

Names of witnesses (if any):_____

Where, exactly, did it happen?_____

What were you doing at the time?_____

Describe step by step what led up to the injury/near miss. (continue on the back if necessary):
_____

What could have been done to prevent this injury/near miss?_____

What parts of your body were injured? _____

If a near miss, how could you have been hurt?_____

Did you see a doctor about this injury/illness? □ Yes □ No

If yes, whom did you see? _____Doctor's phone number:_____

Date: _____Time:_____

Has this part of your body been injured before? □ Yes □ No

If yes, when? _____Supervisor:_____

Your signature:_____ Date:_____

Figure 4.3 Example of OSHA incident notification form.

Supervisor's Accident Investigation Form

Name of Injured Person _____

Date of Birth _____ Telephone Number _____

Address _____

City _____ State_____ Zip _____

(Circle one) Male Female

What part of the body was injured? Describe in detail. _____
_____

What was the nature of the injury? Describe in detail. _____
_____
_____

Describe fully how the accident happened? What was employee doing prior to the event? What equipment, tools being using? _____
_____
_____

Names of all witnesses:
_____   _____
_____   _____

Date of Event _____ Time of Event _____

Exact location of event: _____

What caused the event? _____
_____
_____

Were safety regulations in place and used? If not, what was wrong? _____
_____

Employee went to doctor/hospital? Doctor's Name _____

Hospital Name _____

Recommended preventive action to take in the future to prevent reoccurrence.
_____
_____
_____
_____

Supervisor Signature _____Date_____

Figure 4.3 Cont'd.

Incident Investigation Report

Instructions: Complete this form as soon as possible after an incident that results in serious injury or illness.

(Optional: Use to investigate a minor injury or near miss that *could have resulted in a serious injury or illness*.)

This is a report of a: ☐ Death ☐ Lost Time ☐ Dr. Visit Only ☐ First Aid Only ☐ Near Miss

Date of incident: This report is made by: ☐ Employee ☐ Supervisor ☐ Team ☐ Other_____

Name: Sex: ☐ Male ☐ Female Age:

Department: Job title at time of incident:

This employee works:

    ☐ Regular full time

    ☐ Regular part time

    ☐ Seasonal

    ☐ Temporary

Months with this employer_____

Months doing this job:_____

Part of body affected: (shade all that apply)

Nature of injury: (most serious one)

    ☐ Abrasion, scrapes

    ☐ Amputation

    ☐ Broken bone

    ☐ Bruise

    ☐ Burn (heat)

    ☐ Burn (chemical)

    ☐ Concussion (to the head)

    ☐ Crushing Injury

    ☐ Cut, laceration, puncture

    ☐ Hernia

    ☐ Illness

    ☐ Sprain, strain

    ☐ Damage to a body system:

    ☐ Other _____

Exact location of the incident: Exact time:

What part of employee's workday? ☐ Entering or leaving work ☐ Doing normal work activities ☐ During meal period ☐ During break ☐ Working overtime ☐ Other_____

Names of witnesses (if any):

4

Figure 4.3 Cont'd.

Number of attachments:

Written witness statements: Photographs: Maps / drawings:_____

What personal protective equipment was being used (if any)?_____

Describe, step-by-step the events that led up to the injury. Include names of any machines,

parts, objects, tools, materials and other important details._____

Description continued on attached sheets: ☐

Unsafe workplace conditions: (Check all that apply)

    ☐ Inadequate guard
    ☐ Unguarded hazard
    ☐ Safety device is defective
    ☐ Tool or equipment defective
    ☐ Workstation layout is hazardous
    ☐ Unsafe lighting
    ☐ Unsafe ventilation
    ☐ Lack of needed personal protective equipment
    ☐ Lack of appropriate equipment / tools
    ☐ Unsafe clothing
    ☐ No training or insufficient training
    ☐ Other: _____

Unsafe acts by people: (Check all that apply)

    ☐ Operating without permission
    ☐ Operating at unsafe speed
    ☐ Servicing equipment that has power to it
    ☐ Making a safety device inoperative
    ☐ Using defective equipment
    ☐ Using equipment in an unapproved way
    ☐ Unsafe lifting
    ☐ Taking an unsafe position or posture
    ☐ Distraction, teasing, horseplay
    ☐ Failure to wear personal protective equipment
    ☐ Failure to use the available equipment / tools
    ☐ Other: _____

Why did the unsafe conditions exist?_____

Why did the unsafe acts occur?_____

Figure 4.3 Cont'd.

Is there a reward (such as "the job can be done more quickly", or "the product is less likely to be damaged") that may have encouraged the unsafe conditions or acts? ☐ Yes ☐ No

If yes, describe:_____

Were the unsafe acts or conditions reported prior to the incident? ☐ Yes ☐ No

Have there been similar incidents or near misses prior to this one? ☐ Yes ☐ No

What changes do you suggest to prevent this incident/near miss from happening again?

☐ Stop this activity ☐ Guard the hazard ☐ Train the employee(s) ☐ Train the supervisor(s)

☐ Redesign task steps ☐ Redesign work station ☐ Write a new policy/rule

☐ Enforce existing policy ☐ Routinely inspect for the hazard ☐ Personal Protective Equipment

☐ Other: _____

What should be (or has been) done to carry out the suggestion(s) checked above?

Description continued on attached sheets: ☐

Written by:_____

Department:_____

Title:_____

Date:_____

Names of investigation team members:_____

Reviewed by: Title:_____

Date:_____

Figure 4.3 Cont'd.

Employees in the process area where the incident occurred should be consulted, interviewed or made a member of the team. Their knowledge of the events represents a significant set of facts about the incident that occurred. The report, its findings, and recommendations should be shared with those who can benefit from the information. The cooperation of employees is essential to an effective incident investigation. The focus of the investigation should be to obtain facts and not to place blame. The team and the investigative process should clearly deal with all involved individuals in a fair, open, and consistent manner.

Emergency Preparedness

Every company needs to identify and plan for the actions necessary for employees to take when there is an unwanted release of highly hazardous chemicals. Emergency preparedness is the company's third line of defense that will be relied on along with the second line of defense, which is to control the release of chemical. Control releases and emergency preparedness will take place when the first line of defense to operate and maintain the process and contain the chemicals fails to stop the release. In preparing for an emergency chemical release, companies will need to decide the following:

- Whether they want employees to handle and stop small or minor incidental releases;
- Whether they wish to mobilize the available resources at the plant and have them brought to bear on a more significant release;
- Whether employees are to evacuate the danger area and promptly escape to a preplanned safe zone area, and then allow the local community emergency response organizations to handle the release; or
- Whether they are to use all or just some combination of these actions.

Companies need to select how many different emergency preparedness or third lines of defense they plan to have, develop the necessary emergency plans and procedures, appropriately train employees in their emergency duties and responsibilities, and then implement these lines of defense.

Companies must have an emergency action plan that will facilitate the prompt evacuation of employees when there is an unwanted release of a highly hazardous chemical. This means that the plan will be activated by an alarm system to alert employees when to evacuate, and that employees who are physically impaired will have the necessary support and assistance to get them to a safe zone. The intent of these requirements is to alert and move employees quickly to a safe zone. Delaying alarms or confusing alarms are to be avoided. The use of process control centers or buildings as safe areas is discouraged. Recent catastrophes indicate that lives are lost in these structures because of their location and because they are not necessarily designed to withstand overpressures from shock waves resulting from explosions in the process area.

When there are unwanted incidental releases of highly hazardous chemicals in the process area, the company must inform employees of the actions/procedures to take. If the company wants employees to evacuate the area, then the emergency action plan will be activated. For outdoor processes, where wind direction is important for selecting the safe route to a refuge area, the companies should place a wind direction indicator, such as a wind sock or pennant, at the highest point visible throughout the process area. Employees can move upwind of the release to gain safe access to a refuge area by knowing the wind direction.

If the company wants specific employees in the release area to control or stop the minor emergency or incidental release, these actions must be planned in advance and procedures developed and implemented. Handling incidental releases for minor emergencies in the process area must include pre-planning, providing appropriate equipment for the hazards, and conducting training for those employees who will perform the emergency work before they respond to handle an actual release. The company's training program, including the Hazard Communication standard training, is to address, identify, and meet the training needs for employees who are expected to handle incidental or minor releases.

Preplanning for more serious releases is an important element in the company's line of defense. When a serious release of a highly hazardous chemical occurs, the company, through preplanning, will have determined in advance what actions employees are to take. The evacuation of the immediate release area and other areas, as necessary, would be accomplished under the emergency action plan. If the company wishes to use plant personnel—such as a fire brigade, spill control team, a hazardous materials team—or employees to render aid to those in the immediate release area and to control or mitigate the incident, refer to OSHA's hazardous waste operations and emergency response (HAZWOPER) standard (Title 79CFR Part 1910.1 20). If outside assistance is necessary, such as through mutual aid agreements between companies and local government emergency response organizations, these emergency responders are also covered by HAZWOPER. The safety and health protection required for emergency responders is the responsibility of their company's and of the on-scene incident commander.

Responders may be working under very hazardous conditions; therefore, the objective is to have them competently led by an on-scene incident commander and the commander's staff, properly equipped to do their assigned work safely, and fully trained to carry out their duties safely before they respond to an emergency. Drills, training exercises, or simulations with the local community emergency response planners and responder organizations is one means to obtain better preparedness. This close cooperation and coordination between plant and local community emergency preparedness managers also will aid the company in complying with the Environmental Protection Agency's Risk Management Plan criteria.

An effective way for medium to large facilities to enhance coordination and communication during emergencies within the plant and with local community organizations is by establishing and equipping an emergency control center. The emergency control center should be located in a safe zone so that it could be occupied throughout the duration of an emergency. The center should serve as the major communications link between the on-scene incident commander and plant or corporate management as well as with local community officials. The communications equipment in the emergency control center should include a network to receive and transmit information by telephone, radio, or other means. It is important to have a backup communications network in case of power failure or if one communication means fails. The center also should be equipped with the plant layout; community maps; utility drawings, including water for fire extinguishing; emergency lighting; appropriate reference materials such as a government agency notification list, company personnel phone list, SARA (Superfund Amendments Reauthorization Act) Title III reports and MSDS's, emergency plans and procedures manual; a listing the location of emergency response equipment and mutual aid information; and access to meteorological data and any dispersion modeling data.

Compliance Audits

An audit is a technique used to gather sufficient facts and information, including statistical information, to verify compliance with standards. Companies must select a trained individual or assemble a trained team to audit the process safety management system and program. A small process or plant may need only one knowledgeable person to conduct an audit. The audit includes an evaluation of the design and effectiveness of the process safety management system and a field inspection of the safety and health conditions and practices to verify that the company's systems are effectively implemented. The audit should be conducted or led by a person knowledgeable in audit techniques who is impartial towards the facility or area being audited. The essential elements of an audit program include planning, staffing, conducting the audit, evaluating hazards and deficiencies and taking corrective action, performing a follow-up and documenting actions taken.

Planning

Planning is essential to the success of the auditing process. During planning, auditors should select a sufficient number of processes to give a high degree of confidence that the audit reflects the overall level of compliance with the standard. Each company must establish the format, staffing, scheduling, and verification methods before conducting the audit. The format should be designed to provide the lead auditor with a procedure or checklist that details the requirements of each section of the standard. The names of the audit team members should be listed as part of the format as well. The checklist, if properly designed, could serve as the verification sheet that provides the auditor with the necessary information to expedite the review of the program and ensure that all requirements of the standard are met. This verification sheet format could also identify those elements that will require an evaluation or a response to correct deficiencies. This sheet also could be used for developing the follow-up and documentation requirements.

Staffing

The selection of effective audit team members is critical to the success of the program. Team members should be chosen for their experience, knowledge, and training and should be familiar with the processes and auditing techniques, practices, and procedures. The size of the team will vary depending on the size and complexity of the process under consideration. For a large, complex, highly instrumented plant, it may be desirable to have team members with expertise in process engineering and design; process chemistry; instrumentation and computer controls; electrical hazards and classifications; safety and health disciplines; maintenance; emergency preparedness; warehousing or shipping; and process safety auditing. The team may use part-time members to provide the expertise required and to compare what is actually done or followed with the written PSM program.

Conducting the Audit

An effective audit includes a review of the relevant documentation and process safety information, inspection of the physical facilities, and interviews with all levels of plant personnel. Utilizing the audit procedure and checklist developed in the preplanning stage, the audit team can systematically analyze compliance with the provisions of the standard and any other corporate policies that are relevant. For example, the audit team will review all aspects of the training program as part of the overall audit. The team will review the written training program for adequacy of content, frequency of training, effectiveness of training in terms of its goals and objectives as well as to how it fits into meeting the standard's requirements. Through interviews, the team can determine employees' knowledge and awareness of the safety procedures, duties, rules, and emergency response assignments. During the inspection, the team can observe actual practices such as safety and health policies, procedures, and work authorization practices. This approach enables the team to identify deficiencies and determine where corrective actions or improvements are necessary.

Evaluation and Corrective Action

The audit team, through its systematic analysis, should document areas that require corrective action as well as where the process safety management system is effective. This provides a record of the audit procedures and findings and serves as a baseline of operation data for future audits. It will assist in determining changes or trends in future audits.

Corrective action is one of the most important parts of the audit and includes identifying deficiencies, and planning, following-up, and documenting the corrections. The corrective action process normally begins with a management review of the audit findings. The purpose of this review is to determine what actions are appropriate, and to establish priorities, timetables, resource allocations and requirements, and responsibilities. In some cases, corrective action may involve a simple change in procedures or a minor maintenance effort to remedy the problem. Management of change procedures need to be used, as appropriate, even for a seemingly minor change. Many of the deficiencies can be acted on promptly, while some may require engineering studies or more detailed review of actual procedures and practices. There may be instances where no action is necessary; this is a valid response to an audit finding. All actions taken, including an explanation when no action is taken on a finding, need to be documented.

The company must ensure that each deficiency identified is addressed, the corrective action to be taken is noted, and the responsible audit person or team is properly documented. To control the corrective action process, the company should consider the use of a tracking system. This tracking system might include periodic status reports shared with affected levels of management, specific reports such as completion of an engineering study, and a final implementation report to provide closure for audit findings that have been through management of change, if appropriate, and then shared with affected employees and management. This type of tracking system provides the company with the status of the corrective action. It also provides the documentation required to verify that appropriate corrective actions were taken on deficiencies identified in the audit.

Better Management

Ten Things a Leader Can Do to Build an OE Culture

1. Know the people, facility and always understand what is going on within your operations and what can potentially affect it. Engage in dialogue with members of the workforce (employees and contractors); inquire about their work and working conditions. Understand and recognize the value of each individual's contribution to incident-free operations and its facility's productivity.

2. Positively reinforce safe behaviors on the spot. Act immediately to mitigate unsafe or environmentally unsound conditions. Share personal examples of safety learnings and observations from both on and off-the-job.
3. Never ignore a suggestion to improve operations.
4. Devote required resources, including your time, to operational excellence. Know your OE network representatives and participate in OE network activities.

5. Sponsor and participate in critical OE processes; make safety observations, participate in a job safety analysis (JSA), see Figure 4.4 (below), or an incident investigation to determine root causes.
6. Set clear, specific, measurable objectives for operational excellence. Communicate frequently with all members of the workforce on the objectives, measures, plans, and progress. Regularly recognize progress on indicators and achievement of results.
7. Role model company standards by always following rules, standards, and safe work practices, holding others accountable for similarly following them and recognizing those that do.
8. Conduct frequent field visits, ask questions about safety, environmental, and reliability conditions and provide immediate pin-pointed feedback (both positive and constructive).
9. Hold yourself and others accountable for operational excellence performance. Include OE performance in ranking, salary, and job selections.
10. Set high, specific standards for continuous improvement of critical OE processes. Share lessons learned and seek out and adopt processes that could improve performance.

Job Safety Analysis (JSA) Worksheet

Date: _____
Department Name: _____
Division/Unit Name: _____ Facility: _____
Job Title/Description:_____

_____

Name(s) of Individuals Performing JSA : _____ _____ _____ __

Critical Steps	Hazards	Controls and/or Recommended Actions

Figure 4.4 Job safety analysis (JSA) worksheet example.

OE Objectives and Structure

Keywords

Accountability; Attitude; Attributes; Communication; Continuous learning; Dependent; DuPont Bradley Curve; HSE Cultural Ladder; Information; Interdependent; Job safety analysis (JSA); Leadership; Objectives; Reactive; Responsibility; Strategy; Team approach; Traits; Values; Vision; World-class.

Developing a global operational excellence (OE) information management system is a long-term commitment. Much like the enterprise-wide financial information systems that most large companies have installed over the last few decades, it represents the work of many years, not months, and is a very large undertaking; and because of the nature of the OE objectives—*there are no shortcuts.*

In almost every single case, developing an OEMS is a matter of building on what a company already has in place. Even if a management team does not yet think in terms of business processes—like every company out there doing business—it **has** a way of managing things and addressing the unexpected. Instituting an enterprise-wide management system means expanding that effort to the corporate scale.

Where a company starts to develop the OEMS depends on the maturity of its existing system(s). Most asset and capital-intensive companies have developed information systems as management system standards have emerged: environmental performance-management systems follow the ISO 14000 standard, quality follows ISO 9000, energy management follows ISO 50001 and API 9100A&B, and risk management follows ISO 31000. Although the guidelines from the International Organization for Standardization and other standards bodies have allowed companies to standardize procedures, many information management systems have evolved from a collection of simple word-processing documents and spreadsheets. But they were rather limited in their ability to provide strategic insights.

These were essentially tactical compliance solutions as opposed to the performance orientation of current OE management systems and they worked well when decision making was limited to one area of operations and one facility. But as sustainable operational performance improvements became more important to communicate business success to investors and customers, so did the need for a strategic enterprise-wide approach to global decision making.

As management systems evolved, companies started analyzing the aggregated data compiled across operations. Many would find information within distinctly different silos within each operating unit and there would be pockets of excellence rather than uniformity across the organization. Examples of these silos of information include systems addressing environmental, equipment reliability, incident, supply-chain, and risk-assessment information—and these would often be found with little linkage between them and viewed as distinctly separate approaches toward managing them as individual objectives.

Applied Operational Excellence for the Oil, Gas, and Process Industries. http://dx.doi.org/10.1016/B978-0-12-802788-2.00005-1
Copyright © 2015 Elsevier Inc. All rights reserved.

Today, information technology is driving advances in companies' ability to aggregate information and data across all these silos to help create integrated management systems. Although corporate executives and their boards of directors are now able to compare operational performance across a variety of operational metrics, efforts continue to collate, compile, and prioritize the information. Some of this is done through the establishments of dashboards and KPIs and this is just one manner in which executives raise the bar on performance.

Many organizations will find that their membership in various industry associations comes with certain expectations for participation in the development of industry standards as well as adherence to these standards. For example, in addition to the legal and regulatory obligations placed on those operating in industry according to their specific locales around the world, petroleum industry companies that have membership in the American Petroleum Institute (API) agree to abide by industry consensus standards such as API 9100. Member companies of The American Chemistry Council (ACC) agree as a condition of membership to uphold and support the ACC Responsible Care initiative and accordingly pledge to operate their business according to the Guiding Principles.

Here is an excerpt from that pledge:

Chemistry is essential to the products and services that help make our lives safer, healthier, and better. Through the Responsible Care initiative and the Responsible Care Global Charter our industry has made a worldwide commitment to improve our environmental, health, safety and security performance. Accordingly, we believe and subscribe to the following principles:

- To lead our companies in ethical ways that increasingly benefit society, the economy, and the environment.
- To design and develop products that can be manufactured, transported, used and disposed of or recycled safely.
- To work with customers, carriers, suppliers, distributors and contractors to foster the safe and secure use, transport and disposal of chemicals and provide hazard and risk information that can be accessed and applied in their operations and products.
- To design and operate our facilities in a safe, secure and environmentally sound manner.
- To instill a culture throughout all levels of our organizations to continually identify, reduce and manage process safety risks.
- To promote pollution prevention, minimization of waste and conservation of energy and other critical resources at every stage of the life cycle of our products.
- To cooperate with governments at all levels and organizations in the development of effective and efficient safety, health, environmental and security laws, regulations and standards.
- To support education and research on the health, safety, environmental effects and security of our products and processes.
- To communicate product, service and process risks to our stakeholders and listen to and consider their perspectives.
- To make continual progress towards our goal of no accidents, injuries or harm to human health and the environment from our products and operations and openly report our health, safety, environmental and security performance.
- To seek continual improvement in our integrated Responsible Care Management System® to address environmental, health, safety and security performance.
- To promote Responsible Care® by encouraging and assisting others to adhere to these Guiding Principles.

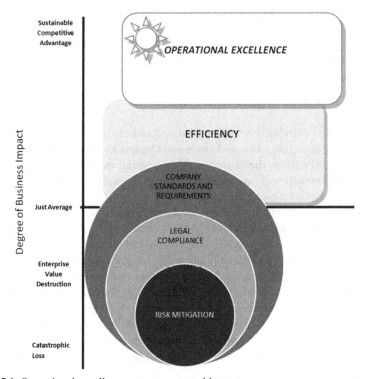

Figure 5.1 Operational excellence components and impacts.

As companies structure and organize their own respective OE management systems they are likely to seek to maximize their own efficiency and effectiveness so as to systematically increase their performance levels through the application of continuous improvement. The manner in which they choose to structure and organize will determine the performance objectives they set for themselves and these tend to be progressively more beneficial as the organization moves incrementally higher as illustrated in Figure 5.1.

Processes and Practices

Formalized OE processes, practices, and programs need to have specialized attributes and characteristics to ensure efficient, reliable and sustainable delivery of the business objectives and performance. Specific OE attributes and characteristics usually need to include the following:

- Clear and well defined, documented, leadership-designated ownership roles, responsibilities and accountabilities.
- Clear and well defined, documented organizational business purpose, scope, objectives, performance standards, and alignment with organizational business strategy.
- Clear and well defined relationships and interfaces with related processes or practices and systems.

- Clear and well defined documented activities, sequences, workflows, and/or procedures/practices/guidelines.
- Methods, tools, technologies and systems that support the OE system.
- Organizational capabilities (e.g., knowledge, skills, abilities, competencies, and capacities to support the OE system).
- Clear and well defined provisions, practices and protocols for process performance measurement, management and continuous improvement.

Processes are interrelated activities designed and integrated to achieve a specific business performance objective and outcome. Organizations use these processes to execute work and deliver the business objectives and outcomes. Practices are the features or characteristics of processes that provide or enable specific targeted performance improvements or capabilities.

The deliberate identification, standardization, enablement, optimization and continuous improvement of an organization's performance critical processes and practices helps to achieve the OE system appropriate for the organization.

Safety and OE achievement requires nothing less than a team approach. All parties to the operation must participate and contribute. Without team commitment, cohesiveness, and accountability extended down through the line management chain of command, objectives will not be met. Senior management's leadership role is perhaps the most important aspect of achieving meaningful strides forward in the OE journey. The tone at the top of the organization is key and if senior management does not endorse or actively demonstrate the strategic and tactical importance of safety it will not be made an integral part of the corporate culture.

Senior management teams are often located many miles away in corporate offices, away from the actual plant site. Can they really be that important in providing effective fire and explosion safety measures at any facility or operation or otherwise ensure OE is fully integrated into the day-to-day activities at the facility level? Absolutely! Senior management sets the tone at the top of the organization and extends responsibility and accountability down through the chain of command and this is key to establishing the proper mindset and attitude of the entire management team towards safety. Actions speak louder than words, and the leadership demonstrated by every member of the management team will signal the amount of importance placed on achieving qualitative or quantifiable safety results. Providing a permissive attitude of leaving safety requirements to subordinates or to the loss prevention personnel will not be conductive or lead to good results. The effect of indifference or lack of concern to safety measures is always reflected top down in any organizational structure and this is what develops over time into the company culture. Executive management must express and contribute to an effective safety program in order for satisfactory results to be achieved. All incidents should be thought of as preventable. Incident prevention and elimination should be considered as an ultimate goal of any organization. Setting arbitrary annual incident recordability limits for incidents may be interpreted by some as allowing some incidents to occur. Where a safety culture is "nurtured," continual economic benefits are usually derived. On the other hand, it has been stated that of the 150 largest petroleum and chemical incidents in the last several decades, they have involved breakdowns in the management of process safety and a lack of organization safety culture which could have prevented these occurrences.

Achieving a World-Class Organizational Safety Culture

There are several models that characterize safety culture within an organization. The two most widely known are the DuPont Bradley Curve and the Hudson/Parker HSE Culture Ladder.

The DuPont Bradley Curve (see Figure 5.2) highlights how to achieve world-class safety performance through applying a management approach to improving safety culture. In a mature safety culture, safety is realized as sustainable, with injury rates

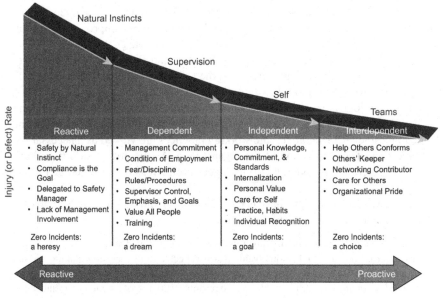

Figure 5.2 DuPont Bradley Curve.

approaching zero. Individuals are empowered to take action as needed to work safely. They support and challenge each other. Decisions are made at the appropriate level and people live by those decisions. The organization, as a whole, realizes significant business benefits in higher quality, greater productivity, and increased profits.

The four stages are further described below:

Reactive Stage

People do not take responsibility. They believe that safety is more a matter of luck than management, and that "accidents will happen." And over time, they do.

Dependent Stage

People see safety as a matter of following rules that someone else makes. Accident rates decrease and management believes that safety could be managed "if only people would follow the rules."

Independent Stage

Individuals take responsibility for themselves. People believe that safety is personal, and that they can make a difference with their own actions. This reduces accidents further.

Interdependent Stage

Teams of employees feel ownership for safety, and take responsibility for themselves and others. People do not accept low standards and risk-taking. They actively converse with others to understand their point of view. They believe true improvement can only be achieved as a group, and that zero injuries is an attainable goal.

A similar arrangement is provided by Hudson and Parker in the 5 step "HSE Culture Ladder" which is characterized by the levels indicated in Table 5.1 and also in Figure 5.3 from the Energy Institute.

At the pathological level an organization displays a failure and lack of willingness to recognize and/or address those issues which may result in poor safety performance. At the highest level, Generative, safe working practices are viewed as a necessary and desirable part of any operation of the organization. As the progression from Pathological to Generative is undertaken, employees are increasing informed and there is increasing trust.

The following generic traits have been recognized in industry as necessary to establish and maintain an effective safety culture.

- *Leadership Safety Values and Actions*: Loss prevention is treated as a complex and systematic concern. It is an organizational value that is reflected in the decision making and daily operations of an organization in managing risk and preventing incidents.
- *Personal Accountability*: All individuals take responsibility for safety and contribute to overall organizational safety.

Table 5.1 **HSE Culture Ladder**

Ladder Step Identifier	Characteristics
Generative	• Safety is integral to how business is handled • Continuous improvement to the organization • Safety viewed as providing profit to the company • New safety ideas and suggestions are encouraged
Proactive	• We work on the issues that we still find • Resources are available to correct issues before an incident • Management is concerned but safety statistics are very important • Procedures are owned by the workers
Calculative	• We have systems in place to manage all concerns and hazards • Numerous safety audits • HSE individuals handling much safety statistics
Reactive	• We do a lot every time there is an incident • Safety is important • We are serious, but why don't they do what they are directed to? • Considerable discussion to re-classify incidents • Safety is very critical after an incident
Pathological	• Who cares? As long as we are not found out! • Our lawyers said it was acceptable • Of course we have incidents, this business is risky • Fire the idiot who had an incident!

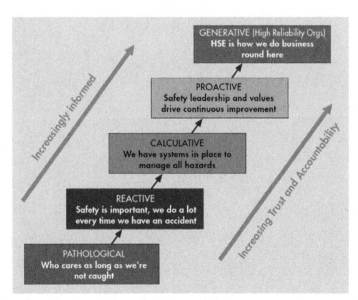

Figure 5.3 HSE Cultural Ladder graphically portrayed.

- *Work Processes*: The process of planning and controlling work activities is undertaken so that system safety is maintained. Concerns of the highest risk will be provided with the most review and management approval.
- *Continuous Learning*: Opportunities to learn risk prevention techniques are continuously researched and implemented by the organization and personnel as appropriate. Hazards, procedures, and job responsibilities are explained and understood by personnel. Safety culture is flexible so that personnel are able to identify and respond appropriately to hazard indications.
- *Environment for Raising Concerns*: A safety-conscious work environment is nurtured, where personnel are allowed to raise safety issues without fear of retaliation, intimation, harassment, or discrimination. They perceive reporting of safety issues as useful to the organization and underreporting is avoided.
- *Effective Safety Communication*: Communication maintains the safety focus and appreciation. Knowledge and lesson learned are shared across the organization, especially where various entities are involved in different phases of a project or activity.
- *Respectful Work Environment*: Trust and respect are reflected throughout the organization.
- *Questioning Attitude*: Individuals avoid complacency and continuously question existing environments and activities to identify concerns that may result in an unsafe condition. A subordinate does not hesitate to question a supervisor, and a contractor employee does not hesitate to question an employee of an operating company.

OE Typical Elements

6

Keywords

Accountability; Asset management; Community; Compliance; Construction; Consumers; Customers; Design; Efficiency; Emergency response; Environmental impact; Focus areas; Human resources; Improvement; Incident investigation; Innovation; Knowledge sharing; Leadership; Learning; Management of change (MOC); OE elements; Pre-startup safety review (PSSR); Product stewardship; Recruitment; Responsibility; Risk management; Self-development; Safety, health and environment (SHE) aspects; Social reliability; Society; Training.

Typical Operational Excellence Elements

An operational excellence management system (OEMS) framework is typically organized logically beginning with focus areas and then though a set of elements that articulate performance objectives and expectations and in turn through the collection of programs, processes, procedures, and activities in a manner that parallels and is aligned with process safety management (PSM). The major enabling elements may not be the same for all organizations due to their own structure and business activities and organizations may differ in the manner in which they choose to organize and structure them, but this is usually attributable to the manner in which things are communicated. They generally contain many of the same common elements.

The 11 most common elements of operational excellence (OE) programs in top global oil, gas, petrochemical companies encompassing the process industry and other asset-intensive companies are presented in Figure 6.1. These 11 common elements commonly characterize the core of their OEMS and tend to be expressed in various forms for specific emphasis with organizations citing more and some fewer. These include; leadership and accountability, risk management, facilities design and construction, asset integrity and reliability, safe operations, environmental health, incident management, emergency preparedness, supply chain & third party services, product stewardship, and of course continuous improvement. These initiatives are nearly always interconnected with discrete management systems, or frameworks of processes and procedures, that expand and evolve with the business through continuous improvement.

OEMS tend to be based on straightforward concepts and in doing so help to cut through the complexity. Armed with an intimate knowledge of its operations at every level, a company's leadership can focus on addressing the most critical risks before they have a negative material impact on the business. They can cut waste, learn to solve problems, develop a culture that prizes continuous improvement, and perhaps most importantly to the results driven executive—drive the organization to sustainability and creation of certain competitive advantages.

Applied Operational Excellence for the Oil, Gas, and Process Industries. http://dx.doi.org/10.1016/B978-0-12-802788-2.00006-3
Copyright © 2015 Elsevier Inc. All rights reserved.

Common Enabling Elements of Operational Excellence

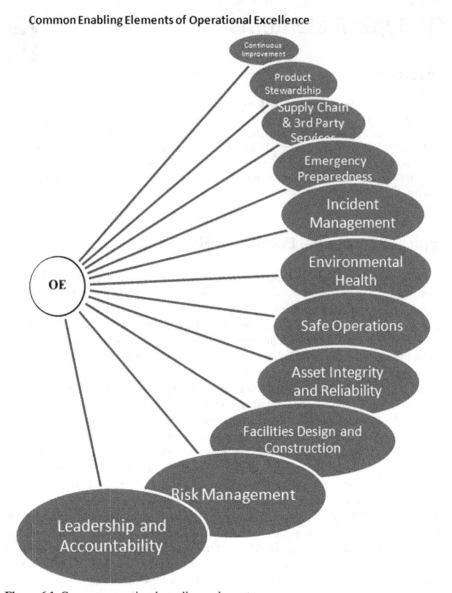

Figure 6.1 Common operational excellence elements.

An enterprise-wide OEMS enables a company to do on an operational basis what well-known enterprise resource planning systems enable it to do on a financial basis. It affords a clear line of sight into corporate activities, from the individual operator level up to the enterprise level. Unlike financial information systems, which track what employees, facilities, and departments are earning or spending, the operational information system tracks *what* they are actually doing and measures *how* they do it.

OEMS are frameworks developed to support the policies and procedures outlined in the OE strategy of the company. These systems underpin that strategy and provide a way to systematically maintain corporate policies and procedures.

So what can companies do with such a system? If a process at a plant in France experiences a failure, the lessons and solutions that emerge from that experience can be quickly deployed to every other similar process the company operates, anywhere in the world. Any new learning, anywhere, can be quickly communicated across the entire company. This type of learning prevents the same unexpected downtime experienced by one facility of the enterprise from recurring in another facility and helps assure with greater certainty that assets will perform as expected.

While this is important, perhaps more so is the way the information conveyed in the OEMS helps to quantify leadership and shape the corporate culture. Managers can measure how, and how quickly, staff deals with a problem. For example, it will help a responsible manager establish how quickly a shift superintendent finds out about an incident or near miss with a potentially high impact; how long it takes to close out all the action items involved in addressing the root causes of the problem. With such information, companies can speed response rates as well as identify and reward employees who respond quickly to high-impact incidents.

A system that focuses on disciplined processes and supports corporate policies and provides greater emphasis to management objectives. Managers can set all the cost-saving goals they like, but it is the operator turning the knobs at the field level who determines whether those goals are met, and at every level of the supervision it is important that there is buy-in and commitment to meeting the company performance goals and objectives. The OEMS may be driven from the top, but the information input and adoption will always be bottom-fed. Operations and maintenance personnel are in the best position to make decisions about what to do in an emergency or how to solve a problem or, even more importantly, to prevent one.

The enterprise-wide information system makes the organization as a whole more agile. Employees at all levels are more conscious of risks they can control and more alert to opportunities to make improvements. Annual or bi-annual audits help gauge certain measures of success, but bottom-up information management enables companies to become more proactive and make faster, more confident decisions that drive incremental improvements in operational efficiency.

Leadership, Management, and Accountability

Management establishes policy, strategy, sets expectations and provides the resources for successful operations. Assurance of operations integrity requires management leadership and commitment visible to the organization, and accountability at all levels.

Expectations related to this element include:

- *Vision and Mission*—Processes are provided to establish, communicate and maintain organizational vision and mission for the organization that inspire, aligns, and sustains activity and behaviors throughout the organization, consistent with the objectives for OE.

- *Commitment to OE*—All levels of management are actively engaged with their respective organizations on issues related to OE system which includes its design, implementation, effectiveness and continuous improvement. These individuals demonstrate clear, consistent and visible commitment to the organizations values and the objectives of OE.
- *Driving OE Culture*—All levels of management ensure that their respective organizations conduct company business with the highest levels of business values and standards. They promote, establish and maintain organizational behaviors and norms consistent with the organizational values and OE objective.
- *Resources*—A process is established that provides for a business strategy to develop long and short range business plans in order to allocate resources consistent with the objective of OE.
- *Customer Focus*—It is important to provide high quality products and/or services to customers within a reasonable time, at a competitive cost and meeting OE objectives. This is achieved through processes that clearly identify customers, evaluate customer requirements, and monitor customer satisfaction in order to improve.

Asset Management (Design, Construction, Operations, Maintenance, Inspection)

Inherent safety and security can be achieved, and risk to health and the environment minimized, by using consistent engineering standards, procedures, and management systems for facility design, construction, operation, maintenance, and inspection

activities. These are usually grouped together in an element identified as Asset Management. This ensures that performance critical assets are designed, fabricated, constructed operated, inspected, and maintained to achieve high levels of reliability, integrity, and efficiency throughout their life cycle. Expectations related to element include the following:

- *Design, Procurement and Construction*—Processes are provided to plan, design, procure and construct reliable and efficient assets in accordance with organizational SHE requirements, company standards, best practices, and applicable governmental standards and regulations.
- *Identification and Assessment of Performance Critical Assets*—Processes are provided for assets that are identified for achieving specific performance objectives have been identified and assessed with respect to critical performance and condition parameters, degradation mechanisms, and failure risks and impacts.
- *Operations*—Processes are provided to ensure assets are operated at their most efficient identified operating modes/envelops in a manner that is cost-effective and compliant with organizational SHE requirements and applicable governmental standards and regulations.
- *Asset Inspection*—Processes are provided for managing potential deterioration of facilities and equipment (e.g., corrosion, fatigue, etc.), testing, and inspection of equipment and facilities to ensure mechanical integrity, reliability and performance over their life-cycle.
- *Maintain Assets*—Processes are provided to proactively, prioritize, plan, schedule, and complete necessary maintenance, for all structures, equipment and protective devices to avoid failures and maximize reliability.
- *Outages, Shutdowns, and Decommissioning*—Processes are provided to ensure outages, shutdowns, and permanent abandonment/removals are properly planned, executed, and managed to maintain the mechanical integrity and value of equipment and facilities when required, and also maintain organizational SHE requirements.
- *Quality Assurance*—Process are provided to undertake quality assurance activities throughout the asset lifecycle in accordance with the company standards, best practices and governmental standards and regulations.
- *Improvements*—Processes are provided to ensure that lessons learned during operation and maintenance and also improvements/upgrades to existing designs, are incorporated into new designs to continuously improve safety, asset integrity, and reliability.
- *Work Procedures*—Processes are provided to ensure that all required work procedures for operating, inspecting, and maintaining assets are established, regularly reviewed and updated, improved, communicated and adhered to.

Risk Management

Risk assessments can reduce safety, health, environmental, security and business risks and mitigate the consequences of incidents by providing essential information for decision-making. Risk is a future event with the ability to impact the organization's mission, strategic and business plans, projects, routine operations, processes, objectives, assets, reputation, or the delivery of stakeholders expectations. Risk management reduces and controls the impact of undesirable events for all aspects to an organization. Providing prudent risk-based assessment and management proactively addresses and mitigates perceived business threats. Financial resources are efficiently managed to yield direct and significant impact on OE objectives reflecting effective utilization of an organization's financial resources and strengthening its competitive advantage. Successful risk assessment and management requires the following:

- *Risk Identification, Assessment and Preparation*—Processes are provided to identify risks, assess their probabilities and potential consequences, undertake prudent management, develop appropriate adverse environment/incident mitigation plans, and monitor and track implementation with continuous improvement to react to changing business environments.
- *Risk Awareness*—Processes are provided to communicate risk and their associated mitigation plans with the organizations, contractors, suppliers and customers as appropriate to improve preparedness, collective and individual decision making and crisis management arrangements.
- *Risk Maturity*—The organization will work towards meeting its minim risk maturity requirements as defined by its maturity matrix/model.
- *Business Continuity Management*—Processes are provided to develop, communicate and implement plans for responding to incidents and unplanned events. It should include identifying qualified personnel and adequate resources required to manage potential adversities.
- *Risk Documentation*—Processes are provided to ensure all documentation related to the risk identification, assessment and management is complete, current, and readily available.

Management of Change

Facilities are continually subject to changes to increase efficiencies, improve operability and safety, accommodate technical innovations, and implement mechanical improvements. Additionally, on occasion, temporary repairs, connections, bypasses, or other modifications may be made due to operating necessities. Any of these changes in operations, procedures, site standards, materials, equipment, facilities, or organizations can introduce new hazards and must be evaluated and managed to ensure that risks arising from these changes remain at an acceptable level.

The management of change (MOC) is typically not applied to changes in specific individuals, but to the organizational structure or significant changes in operating conditions. For example, if the number of individuals in a particular area is changed or there are significant changes in operating conditions for employees, e.g., a change from a 8 h shift to a 12 h changes must be evaluated prior to their implementation. A like for like (i.e., identical) replacement is not normally considered a MOC event. A MOC procedure must include the technical basis/justification for the change, an assessment of the safety, health, and environmental impacts (i.e., hazard analysis) and if necessary any mitigation measures to accept the new risk level posed by the change, time period necessary for the change (temporary or permanent, and implementation period) and management approval of the change. If the change results in modification to operating procedures, these also must be implemented before the change is activated and personnel must be informed and if necessary trained in the new procedures. Temporary changes must indicate the durations and specific dates of application. The required authorizations for the change to be implemented must be secured before it takes effect. An example of a Management of Change form for a process facility is provided in Figure 6.2.

Pre-startup Safety Reviews

The term pre-startup safety review first gained prominence in the process industries with the introduction of the PSM regulations in the United States (OSHA and EPA). The purpose of the pre-startup safety reviews (PSSR) is to ensure that any changes made to a facility or process meet the design or operating intent. The PSSR aims to find any changes that may have occurred during detailed design engineering or during the construction phase of the project. The PSSR not only covers equipment, but also "soft" issues such as training and operating procedures. The PSSR plays a particular role in large projects, as these invariably run over budget and schedule, thus creating pressure on the project team to eliminate or delay the installation of items that are not absolute necessary for startup. If not adequately assessed, this can lead to corner cutting, either intentional or non-intentional, which could affect safety aspects. By undertaking the PSSR, operating organizations can refuse to accept a facility that is considered unsafe. It should be noted that the PSSR is not intended to replace the mechanical completion check by project engineers and inspection groups, but to ensure such activities are actually fully accomplished and items resolved. Nor is the PSSR expected to re-evaluate the project hazard reviews or undertake new hazard assessments, but to ensure any recommendations from these reviews have been addressed and resolved. Figure 6.3 provides an example of a PSSR form that is commonly utilized in the process industries.

Human Resources and Training

Control of operations depends upon people. Achieving OE requires the appropriate screening, selection, placement, continuous assessment, and training of employees and assuring qualified and well informed contractors. A training process should contain the type, method, length, frequency, and content of the program. Additionally a method to assess the quality and effectiveness of the training needs to be provided which could include audits, written tests, hands-on tests, personnel interviews, surveys, etc. Key processes include

Management of Change						
Section 1 Initiation (To be completed by Originator)						
Originator (Name)		Employee ID:		Dept/Unit		Date / /
Description of Change:						
Does This Change Affect Safety No □ Yes □ Explain:						
Is the Change Permanent □ Temporary □ Temporary Change Expiration Date / /						
Anticipated Completion Date / /Signature:						
Section 2 MOC Request Authorization (Originator's Supervisor, e.g. Maintenance, Operations						
Name:		Employee ID	Signature:		Date / /	Telephone:

Section 3 Authorization Requirements: (To be completed by MOC Coordinator)
Each designated individual must complete and sign his assessment by   /   /

PSM Components	Req'd	Reviewer	No	Yes*	Signature
Technical Basis of Process	□		□	□	
Utilities	□		□	□	
Piping	□		□	□	
Electrical and Power	□		□	□	
Mechanical Fixed	□		□	□	
Mechanical Rotating	□		□	□	
Instr., ESD and DCS	□		□	□	
Safety/Fire Protection	□		□	□	
Industrial Hygiene/Medical	□		□	□	
Environmental	□		□	□	
Emergency Response	□		□	□	
PrHA	□		□	□	
PHA/HAZOP/What If/QRA	□		□	□	
Operating Procedures	□		□	□	
Emergency Procedures	□		□	□	
Mech. Integrity (QA/QC)	□		□	□	
Training	□		□	□	
Workshops/Contractors	□		□	□	
On-the-Job Training (OJT)	□		□	□	
Pre-Startup Safety Review (PSSR)	□		□	□	
PSM Documentation e.g., MSDS	□		□	□	
PFD/P&ID/Data Sheets, etc.	□		□	□	
Fixed Equipment Files	□		□	□	
Rotating Equipment Files	□		□	□	
ESD, DCS & Instr. Files	□		□	□	

Section 4 Action Required (by items referenced above*):

Section 5 Document Closure (to be completed by MOC Coordinator and routed to Implementing Organization, e.g., Engineering, Operations, etc.)
□ Approved □ Rejected □ Training Required
MOC Coordinator Signature   Document Closure Date

Figure 6.2 Example of management of change (MOC) form for a process facility change.

Pre-Startup Safety Review Form				
Facility				
Section 1 Project Description (To be completed by Project Engineer or Facility Manager)				
Process	New Facility ☐		Modification ☐	
Project Description:				
Section 2 Checklist (To be completed by Project Engineer or Facility Manager)				
The construction and equipment has been checked for conformance with design specifications and applicable codes and standards.			Yes	No
Applicable process safety information (PSI) has been developed or updated as necessary to reflect the new/modified process.			Yes	No
Plant wide safe work procedures are adequate considering the new/modified process conditions.			Yes	No
Standard operating procedures and emergency procedures have been developed or updated as needed to ensure safe operation of the new/modified process.			Yes	No
Maintenance procedure and a preventative maintenance schedule have been developed or updated as needed to ensure safe operation of the new/modified process			Yes	No
For new processes, the modification has been subjected to management of change review and all recommendations have been resolved or implemented before startup.			Yes	No
For modified processes, the modification has been subjected to management of change (MOC) review and all recommendations have been resolved or implemented before startup.			Yes	No
Training of each employee involved in operating and/or maintaining the process has been completed and documented.			Yes	No
Section 3 Approvals (Name and Date)				
Project Engineer				
Safety Engineer				
Facility Manager				
Section 4 Deficiencies/Action Items (To be completed only if initially disapproved)				
Deficiencies/Action Item	Assigned To	ETC	Date Resolved	
Section 5 Approvals If Deficiencies Noted (Name and Date)				
Project Engineer				
Safety Engineer				
Facility Manager				

Figure 6.3 Pre-startup safety review example form.

self-development, employee performance assessment/feedback, development & improvement, engagement of employees, employee recognition & reward, and knowledge sharing.

The key processes expectations include the following:

- *Employee Recruitment, Selection and Retention*—Processes are provided that (1) Properly evaluate, select, and place employees based on their qualifications, abilities, and performance, (2) Attract and retain talents and high performance individuals, (3) Properly orientate new employees in their positions and duties, (4) Regularly evaluate employee capabilities and potential, (5) Succession plans are undertaken.
- *Employee Competency Assessment and Development*—Process are provided to assess individual competencies, certification needs, and development plans to avoid skill gaps and enable continuous development toward full employee potentials.
- *Self-Development*—Process are provided to promote, encourage, and enable employee personal self-development which will foster individual ownership and accountability.

- *Employee Engagement and Feedback*—Processes are provided to encourage employees to periodically provide feedback, suggest improvements, and share concerns to enhance satisfaction, motivation and development.
- *Employee Reward and Recognition*—Processes are provided to recognize and reward employees based on their behavior and achievements.
- *Knowledge Sharing*—Processes are provided to facilitate and ensure that knowledge is captures, shared and retained.

Reliability and Efficiency

Identify and resolve facility, business work process and human reliability and efficiency concerns that may cause significant incidents or performance gaps.

Reliability: Optimizing operating success becomes easier when you structure and organize the business as a holistically integrated system where every process and everyone works together to optimize production performance. Reliability increases the chance of success, so a company built as a "system-of-reliability" maximizes its operating profits and production success.

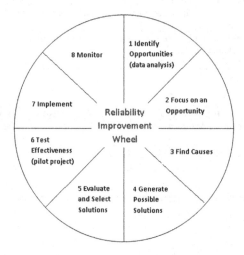

The three requirements needed to create a system-of-reliability for OE are:

- A defect eliminating work quality assurance system where your processes are robust and avoid disruption, and ensure right-first-time results. Only if processes are built to go right every time, avoid errors, and proactively prevent causes of problems, can you reach OE.
- Business-wide process innovation focused on optimizing for the highest productivity, least cost, and right quality output. Then you install the next generation of solutions for ever better productivity.
- Holistic, lifecycle physical asset management for stable, reliable operation with outstanding availability, highest utilization and most sustainable throughput.

Efficiency: This aspect examines improving the efficiency of an organization through better systems, marketing, time management and superior operational structures. Operational efficiency for businesses can be defined as the ratio between the input to run a business operation and the output gained from the business. By improving operational efficiency, the output to input ratio improves. Inputs are typically the money (cost), people (headcount) and time or effort. Outputs are typically the money (revenue, margin, cash), new customers, customer loyalty, market differentiation, headcount productivity, innovation, quality, speed & agility, complexity or opportunities.

When improving operational efficiency, the most common utilized are:

- Same output for less input
- More output for same input
- Much more output for more input

Product Stewardship

Product stewardship is a way that everyone who designs, makes, sells, uses, and recycles or disposes of products can help protect safety, health, and environmental quality. It is a concept that calls on everyone in the product life cycle to share responsibility. It is also known as extended product responsibility, product stewardship calls on those in the *product life cycle*—manufacturers, retailers, users, and disposers—to share responsibility for reducing the environmental impacts of products.

Product stewardship recognizes that product manufacturers must take on new responsibilities to reduce the environmental footprint of their products. However, real change cannot always be achieved by producers acting alone: retailers, consumers, and the existing waste management infrastructure need to help to provide the most workable and cost-effective solutions. Solutions and roles will vary from one product system to another.

In most cases, manufacturers have the greatest ability, and therefore the greatest responsibility, to reduce the environmental impacts of their products. Companies that are accepting the challenge are recognizing that product stewardship also represents a substantial business opportunity. By rethinking their products, their relationships with the supply chain, and the ultimate customer, some manufacturers are dramatically increasing their productivity, reducing costs, fostering product and market innovation, and providing customers with more value at less environmental impact. Reducing use of toxic substances, designing for reuse and recyclability, and creating take back programs are just a few of the many opportunities for companies to become better environmental stewards of their products. Forward-thinking businesses have recognized that demonstrated corporate citizenship and maximum resource productivity are essential components to creating competitive advantage and increasing shareholder wealth.

As the sector with the closest ties to consumers, retailers are one of the gateways to product stewardship. From preferring product providers who offer greater environmental performance, to educating the consumer on how to choose environmentally preferable products, to enabling consumer return of products for recycling, retailers are an integral part of the product stewardship revolution. All products are designed with a consumer in mind. Ultimately, it is the consumer who makes the choice between competing products and who must use and dispose of products responsibly. Without consumer engagement in product stewardship, there is no "closing the loop." Consumers must make responsible buying choices which consider environmental impacts. They must use products safely and efficiently. Finally, they must take the extra steps to recycle products that they no longer need.

Solid waste programs in are primarily managed at the regional and local governmental level. Thus, these authorities are essential to fostering product stewardship, especially as it relates to waste management. A growing number of progressive authorities are incorporating product stewardship objectives into their solid waste master plans, and undertaking cooperative efforts with manufacturers, retailers and others to increase recycling of discarded products. They are developing take back mandates for selected products (especially electronics, but other products are under consideration as well). The governmental procurement officials are also encouraging product stewardship innovations through their purchasing programs.

The US EPA's Product Stewardship program has primarily focused on end-of-life considerations as one means of encouraging more environmentally conscious design and greater resource conservation. However to address the full range of product lifecycle issues, the Product Stewardship program also works with other EPA programs, as well as various public- and private-sector stakeholders, to promote "greener" design, greener product standards, and greener purchasing practices.

Compliance Assurance

Verify conformance with company policy and governmental regulations. Ensure that employees and contractors understand their related responsibilities.

The key processes expectations include the following:

- *Identify Requirements*—Processes are provided to identify applicable legal requirements from governmental laws and regulations, ensure these are in conformance with OE policies and requirements.
- *Communication and Reporting*—Processes are provided to ensure employees are aware, understand, and comply with identified requirements. Employees are encouraged to report existing or potential violations of law or company policies, without fear of retribution or company adverse action.
- *Identify Concerns, Management Advisement, and Resolution*—Processes are provided to identify and report significant non-compliance issues and root causes to management in a timely manner and conduct follow-up corrective actions to resolve the concerns as directed by management.

Emergency Management (Response and Incident Investigation)

Emergency planning and preparedness are essential to ensure that, in the event of an incident, all necessary actions are taken for the protection of the public, the environment and company personnel and assets. Integrated emergency response plans (ERPs) for both plant operational personnel and for offsite emergency services (i.e., fire departments, medical ambulances, police, mutual aid organizations, coast guard, etc.) need to be provided, critiqued, drilled, and assessed periodically. Effective incident investigation, using industry recognized methodologies (e.g., Taproot, 5-Whys, root cause analysis (RCA), Fishbone, etc.), incident reporting and follow-up are necessary to provide the opportunity to learn from reported incidents and to use the information to take corrective action and prevent recurrence from the identified root causes (e.g., lessons learned bulletins), which is communicated to all appropriate employees.

External Services

Contractors and suppliers are managed to ensure that the products and/or services supplied on behalf of the organization are provided in a manner that is consistent with OE objectives, expectations, and standards. The various external services which are provided by contractors, suppliers and various commercial agents directly affect the organization's performance, ability to deliver on commitments, SHE goals, and reputation. The expectations associated external services include the following:

- *Communication*—Processes are provided to communicate the organization's standards, procedures and OE objectives and expectations to external service providers.
- *Performance Assessment*—Processes are provided to periodically access and monitor contractor and supplier performance against OE objectives and expectations.
- *Contracting Strategies*—Processes are provided to develop and implement contracting strategies that mitigate performance risks and improve service quality.
- *Networking*—Processes are provided to establish partnerships with suppliers and contractors to optimize expertise, resources, and knowledge sharing for the achievement of common goals and objectives.

Social Responsibility

Social responsibility is an ethical framework which suggests that an entity, be it an organization or individual, has an obligation to act for the benefit of society at large. This responsibility can be passive, by avoiding engaging in socially harmful acts, or active, by performing activities that directly advance social goals. Therefore, leading organizations have an obligation to work ethically and constructively to influence proposed laws and regulations and debate on emerging issues. These organizations also continuously support the social and economic development of their nearby communities and serve as a role model for others. Evidence suggests that social responsibility taken on and adopted voluntarily by companies will be much more effective than corporate social responsibility mandated by government entities. All companies have a two-point agenda—to improve qualitatively (the management of people and processes) and quantitatively (the impact on society). The second is as important as the first and stake holders of every company are increasingly taking an interest in "the outer circle"—the activities of the company and how these are impacting the environment and society.

These programs, processes, and practices which promote and facilitate social responsibility include the following items:

- Utilization of local resources and business development.
- Protection and preservation of workplace and public health, safety, and the environment.
- Sustainable energy and resource development, conservation, and utilization.
- Volunteerism, community awareness and education, and community outreach and service.

Outreach Activities

Stakeholder Engagements
COMMUNITY MEETINGS

Innovation, Learning and Continuous Improvement

Continuously improve operations and accountability to achieve higher levels of safety culture, technology, management, and overall company performance. Individuals at all levels should collectively, actively and systematically identify, evaluate, develop, and promote adoption of best practices and lessons learned to continually improve performance and increase their capacity to achieve desired results. A successful continuous improvement element requires the following:

- Innovative Culture—Processes are provided for organizational norms and a culture is in place to generate, identify, and develop innovative ideas and technologies to improve performance.
- Root Cause Analysis—Processes are provided to identify, analyze, document, communicate and mitigate or eliminate root causes of performance problems and operational incidents. Industry-recognized RCA techniques should be applied when investigating and identifying lessons learned from incidents, near misses, and performance concerns. Investigation recommendations should be tracked until resolution.
- Benchmarking—Processes are provided to ensure regular and objective internal and external benchmarking (i.e., comparison) of performance measures to identify improvement opportunities.
- Performance Monitoring—Processes are provided to monitor performance using quantitative measurement and/or key performance indicators.
- Learning Organization—Processes are provided to ensure effective and efficient communication, dissemination and implementation of lessons learned, improvement ideas, and best practices throughout the organization.
- Continuous Improvement—Processes are provided to continuously identify improvement opportunities. This is accomplished by reviewing elements and processes against industry best practices, new technologies, innovations, suggestions and lessons learned (industry and internal).

Table 6.1 provides examples of some major process companies OE program elements.

Table 6.1 Examples of Various Company Operational Excellence Program Elements

Company A	Company B	Company C	Company D
Element 1: Security of personnel and assets	Element 1: Management leadership, commitment and accountability	Element 1: Leadership	Element 1: Leadership and accountability
Element 2: Facilities design and construction	Element 2: Risk assessment and management	Element 2: Organization	Element 2: Risk assessment and management
Element 3: Safe operations	Element 3: Facilities design and construction	Element 3: Risk	Element 3: Communications
Element 4: Management of change (MOC)	Element 4: Information/ documentation	Element 4: Procedures	Element 4: Competency and training
Element 5: Reliability and efficiency	Element 5: Personnel and training	Element 5: Assets	Element 5: Asset integrity
Element 6: Third-party services	Element 6: Operations and maintenance	Element 6: Optimization	Element 6: Safe operations
Element 7: Environmental stewardship	Element 7: Management of change (MOC)	Element 7: Privilege to operate	Element 7: Contractors, suppliers and others
Element 8: Product stewardship	Element 8: Third-party services	Element 8: Results	Element 8: Emergency preparedness
Element 9: Incident investigation	Element 9: Incident investigation and analysis		Element 9: Incident reporting and analysis
Element 10: Community awareness and outreach	Element 10: Community awareness and emergency preparedness		Element 10: Community awareness and off the job safety
Element 11: Emergency management	Element 11: Operations integrity assessment and improvement		Element 11: Continuous improvement
Element 12: Compliance assurance			
Element 13: Legislative and regulatory advocacy			

OE Key Processes and Safety, Health and Environment (SHE) Embedding

7

Keywords

Accountability; Benchmarking; Benefits; Budget; Business strategy; Commitment; Communication; Compensation; Continuous improvement; Customer focus; Element expectations; Employee training; Engagement; Framework; Knowledge sharing; Leadership; Motivation; Objectives; Obstacles; Process elements tools; Process model; Recognition; Responsibility; Rewards; Risk assessment; Road map; Self-development; Timeframe.

In establishing an operational excellence (OE) framework, an organization will first decide upon the focus areas and then decide upon the enabling elements that will frame the OE management system. Within each of the enabling elements there will be certain expectations outlined that reflect what each should accomplish at a fairly high level. Then in each of these enabling elements the individual processes, programs, procedures, and practices will be mapped out accordingly within this structure. Although the structure and organization in such a graphical presentation might give the impression that each of these topics are completely independent and perhaps even resembling something of various different silos, the reality is that in actual practice these functions are often very interrelated. For example, the efforts made to assure effective communication processes (e.g., daily tailgate safety meetings or communication turnover of key safety information at shift change) can be of critical importance to the effectiveness of safe operations processes for control of hot work (e.g., clearly communicating with all affected stakeholders on operational issues that might affect their work activities).

When an organization decides to create an operational excellence management system (OEMS) framework and use it in conjunction with a safety, health and environmental management system framework, there is going to be considerable overlap and inter connection between many aspects of nearly all the elements. For example, risk assessment and management, asset integrity, safe operations, incident reporting and investigation, change management, and human resources are integral to both OE and safety management. It is worth noting that the OE management systems that we have viewed of the major energy industry players typically begin with leadership and accountability and typically end with a solid commitment to continuous improvement. We believe this is no coincidence, as without leadership commitment to the OEMS there is little hope for the sort of progress most companies strive for as they pursue their vision for commercial success which usually includes some aspect of operational discipline, well executed and reliable operations, best in class performance in their respective market segment(s), market share, and sustainable competitive advantage. We have mentioned the focus areas and enabling elements

Applied Operational Excellence for the Oil, Gas, and Process Industries. http://dx.doi.org/10.1016/B978-0-12-802788-2.00007-5
Copyright © 2015 Elsevier Inc. All rights reserved.

and the basic expectations, each successive layer providing more detail to describe the framework. We have already profiled and listed examples of enabling elements in Chapter 6 with Table 6.1. The next layer are the management processes, and in turn programs and procedures. The individual processes, programs, and procedures are where the real work in managing the OEMS takes place in terms of controlling the work.

Many consider process management as the central core to OE because this is where most of the work gets done. This is because it directly involves managing resources and influencing people to make decisions and take actions that are good for business. They provide a structured means for measuring and taking proactive measures to continuously improve and are indeed key to achieving many of the performance expectations. There is an adage in quality management that says *"what gets measured gets managed"* and that idea holds true here and actually says a lot about how processes can help the manager structure and organize for results. As noted earlier, when an incident occurs or something goes wrong that triggers incident reporting and investigation efforts, the line organization will typically apply the necessary resources to determine what went wrong and take the steps necessary to promptly take corrective action as well as share and apply lessons learned in order to prevent recurrence. However, waiting for an incident to occur before taking action is *reaction*, not control; we talked earlier about the importance of leadership and extending accountability—taking action before an incident occurs to control the work is a critical responsibility for the line manager and it is essential to establishing and maintaining operational discipline. The areas of safety and loss prevention have long provided such a primary focus on proactive prevention as opposed to just reaction. Many companies that go to the trouble of organizing the broader OE framework naturally include provisions of the safety management system as a platform to expand upon to incorporate other aspects of the OE management system for just this reason. While safety is of importance through an enterprise, not every single topic selected for inclusion in an OE management system will have an urgent and direct impact on the effectiveness of the safety management system which is often seen as a subset of the overall OE management system. We have seen numerous such examples included in energy industry OE management systems that have been selected for inclusion in the organization's enabling elements along with their respective expectations and processes, programs and procedures. These might address areas such as "customer focus," "financial resources," and "external services" to name but a few.

As noted earlier, this is largely due to the emphasis placed on the functional aspects of managing the interrelated functional components of an effective process—*responsibilities, standards, documentation, training as well as measurement, and effectively extending accountability.*

How does an organization determine which management processes need to be included in their OE framework? Below are some considerations for identifying and organizing processes with the OE framework.

The Process Management Road Map

An energy industry process management road map consists of numerous components. These include:

- Identify processes and determine applicability.
- Review/modify as required
  - Assign responsibilities
  - Process support team review. (Executive Sponsor/Process Champion) This governing executive team within the process group is composed of the executive in charge of operations. The process support team helps the process group determine its annual guiding principles and goals. The team and group might meet approximately every 2 months.
  - Identify process owner.
  - Process documentation.
    - Add/modify references
  - Process control framework placement. At the beginning, when new processes were being identified and documented, they needed to be placed in the framework. Because there are fewer new processes today, the framework is mostly used during optimization efforts, where the team drills down and completes additional documentation.
- Identify process measurements (KPIs and metrics).
- Knowledge management. This step addresses what knowledge the process owners and performers need to do their work and where the group will place it so those affected can find it easily.
- Documentation and records management.

Guidance for adopting a process framework with the OEMS.

- Establish or select a process control framework to serve as a foundation.
- Align framework with an existing business model.
- Apply centralized governance of the framework—a good idea for a program like Sarbanes Oxley (SOX) that has to be managed at the corporate level.
- Reinforce usage of the model in all aspects of process management and optimization.
- Create a framework that is transparent to the process owners. It is not necessary to provide all the details about it; keep it simple enough to figure out where to find the documents.
- Integrate process and knowledge management efforts consistent with the business strategy and process model.
- Begin with creating and optimizing business processes, and then focus on knowledge management.
- Treat knowledge management as part of the process and not a separate initiative.
- Align business processes, records, and knowledge management as a natural progression.
- Keep it simple—integrate as it makes sense.
- Ensure that the approach is practical, cost-effective, and nondisruptive to normal business operations.

Listed below are some examples of safety processes, programs, activities that may be included in the framework (note these are not all inclusive but merely a few examples presented to give the reader an idea for the progressive detail in each general area) (Tables 7.1 and 7.2).

Table 7.1 **Examples of Elements**

Enabling Element	Expectations	Process, Programs, Activities
Competency and training	Training needs	Assess/update competency needs
		Assess/update training needs
	Training plan	Develop annual training plan
	New and transferred employees	Verify qualifications
		Verify fitness through medical evaluation
	Effective training	Orient workers
		Deliver comprehensive skills training
	Safety training	Provide initial training
		Conduct management training
		Provide specialist/topical training
		Hold unit safety meetings
		Provide on-the-job training
		Provide off-the-job training
		Communicate/train through pre-job discussions
		Provide driver training
	Refresher training	Provide refresher training
	Training records	Document training
	Review for effectiveness	Review and improve training
Incident reporting and analysis	Incident reporting	Institute incident reporting process
	Incident classification	Classify incidents based on company guidelines
	Investigations	Conduct timely investigations
	Corrective actions	Implement corrective actions
	Communication	Communicate lessons learned from incidents
		Apply lessons learned
	Analysis	Track performance of people/equipment
		Analyze incident reports for trends
	Training	Provide employee training for incident reporting and investigation
		Orient contractors in incident reporting process
	Periodic reviews	Complete annual review of incident reporting and analysis processes
Change management	Change management applied in operations, procedures, person- nel, technology and operations	Assign OC coordinator
		Identify change type (process technol- ogy, material, equipment, procedures, facilities, buildings or organizational)
		Identify urgency of the change
		Assign change review team

Table 7.2 **Matrix of Enabling Elements, Processes Process Owners, and Number of Performance Measures**

Enabling Element	OE Process	Process Owner	Designated Champion	No of Performance Measures
Leadership and accountability	Vision, mission and strategies	Dept. head	J. A. Doe	1
	Resources	Planning & resources	A. N. Other	3
	Commitment to OE and OE culture	Dept. head	J. A. Doe	2
Customer focus	Identify customers, needs and satisfaction	Customer administrator head	A. N. Other	1
Human factors	Employee selection	Support group	J. A. Doe	5
	Employee competency assessment and development	Support group	A. N. Other	7
	Self-development	Support group	J. A. Doe	2
	Employee engagement and feedback	Dept. head	A. N. Other	3
	Employee reward and recognition	Dept. head	J. A. Doe	2
	Employee transfer	Administrator	A. N. Other	2
	Knowledge sharing	Administrator	J. A. Doe	2
	Supervision selection	Dept. head	A. N. Other	3
Asset management	Design, procurement, construction	Dept. head, projects	J. A. Doe	Numerous
	Safe, reliable & efficient operation & maintenance	Dept. head, bldg administration	A. N. Other	Numerous
	Assets monitoring & continuous improvement	Dept. head, bldg administration	A. N. Other	Numerous
Process management	Process management	Administrator	J. A. Doe	1
	Environmental protection	Administrator	A. N. Other	Numerous
	Energy efficiency	Administrator	A. N. Other	Numerous
Financial resources	Efficient resource utilization	Planning & support group	J. A. Doe	1

Continued

Table 7.2 Matrix of Enabling Elements, Processes Process Owners, and Number of Performance Measures—cont'd

Enabling Element	OE Process	Process Owner	Designated Champion	No of Performance Measures
External services	Communication of expectations	Administrator	J. A. Doe	1
	Performance assessment	Administrator	A. N. Other	1
	Contracting strategies	Planning and support group	A. N. Other	1
	Building networks	Administrator	A. N. Other	3
Policies and strategies	Communicate understanding of policies	Corp administrator of communications	J. A. Doe	1
Information and document management	Communication of information	Information administrator	J. A. Doe	1
	Accessibility to information	Information administrator	J. A. Doe	2
	Data archiving and information protection	Information administrator	J. A. Doe	2
Change management	Management of change	Dept. head	J. A. Doe	1
Risk management	Risk management	Dept. head	J. A. Doe	4
	Business continuity management (emergency)	Dept. head	A. N. Other	3
	Business continuity management (contingency)	Dept. head	A. N. Other	2
Innovation, learning and continuous improvement	Innovative idea generation, development and implementation	Administrator	J. A. Doe	5
	Root cause analysis	Administrator	A. N. Other	2
	Benchmarking	Administrator	J. A. Doe	1
	Performance monitoring	Administrator	A. N. Other	Numerous
	Learning organization	Administrator	J. A. Doe	1
	Continuous improvement	Dept. Head	A. N. Other	Numerous
Corporate Social responsibility	Corporate social responsibility	Corp administrator	J. A. Doe	3

Leadership, Commitment and Accountability

Leadership is considered the single factor of success in OE. Organizational OE leaders establish a particular vision and set objectives in order to achieve world class results. By their actions, leaders cascade, manage and drive execution, reinforce the OE culture, instill operational discipline and work to ensure that they and the entire workforce comply with OE requirements. Leaders have to visibly demonstrate their commitment through personal engagement with the workforce by showing concern for the health and safety of every individual. They also must demonstrate the same commitment to process safety and environmental protection. They direct the management processes, setting priorities and monitoring progress on plans that focus on the highest impact items. All leaders, no matter what their organizational role, are therefore accountable for enabling OE performance parameters. Activities and requirements for OE parameters for leadership, commitment and accountability are mentioned, described and detailed in in Chapter 6.

Risk Assessment and Management

The introduction of risk management and ensuring its effectiveness requires strong and sustained commitment by the management of an organization. Both strategic and rigorous planning along with commitment is required at all levels. To achieve this management must do the following:

- Define and endorse the risk management policy.
- Ensure that the organization's culture and risk management policy are aligned.
- Determine the risk management performance indicators which will align with performance indicators of the organization.
- Align risk management objectives with the objectives and strategies of the organization.
- Ensure legal and regulatory compliance.
- Assign accountabilities and responsibilities for risk management functions at appropriate levels within the organization.
- Ensure that the necessary resources are allocated to risk management.
- Communicate the benefits of risk management to all stakeholders.
- Ensure the framework for risk management is continuously evaluated for relevance and improvements.

The organization's risk management policy should adequately address objectives and commitment. It should clearly address the following:

- Organization's rationale for managing risk.
- Links between organizational objectives and policies to the risk management policy.
- Accountability and responsibility for managing risk.
- Methodology for addressing conflicting interest.
- Commitment of resources.
- Commitment to review and improvement.
- Communication of risk management policy.

Risk treatment usually adopts one of the following options:

- Avoiding the risk by not implementing or continuing with the activity that has generated the risk.
- Undertaking the risk in order to pursue an opportunity.
- Eliminating the risk source.
- Changing the likelihood generated by the risk.
- Changing the consequences generated by the risk.
- Sharing the risk with another entity or entities which includes contracts and risk financing.
- Retaining the risk by informed decision.

Risk management should be embedded into the policy development, business and strategic planning and review, and change management processes. Further discussion on risk management activities within OE parameters are mentioned and described in Chapter 6.

OE/SHE Key Performance Indicators (KPIs)

8

Keywords

Communication; Dashboard; Data collection; Effectiveness; Evolution; Hazards; Illness; Improve performance; Incident pyramid; Incidents; Indicator; Injuries; Key performance indicators (KPIs); Lagging; Leading; Limitations; Measure; Near miss; Observations; Parameters; Performance; Practicality; Proactive; Process Safety Event (PSE); Reactive; Reliability; Safety meetings; Target; Tiered KPIs; Tolerances; Tracking; Usefulness; Validity.

You cannot manage what you do not measure. This statement represents an important idea within quality management. The heart of management control is measuring performance in quantifiable, objective terms. Measuring safety performance solely through the use of statistics generated by simply counting incidents after the fact can pose serious limitations and if used in isolation as the only means of safety performance success, it is often one of the most misused measurements of safety performance for an organization over a given period of time. The downside of using these indicators in isolation is that they are subject to many variables and forms of manipulation. Often driven by fear of failure and the need to look good to others, results are easily skewed by the failure to report all incidents properly, arbitrary willingness to overlook certain incidents and other such sleight of hand maneuvers that attempt to present losses in a more favorable light.

Looking at incident rates such as the *lost time injury* rate, the *restricted duty injury* rate, or the *first aid injury* rate primarily serves to indicate how well the organization can report and compile the statistics and helps serve as a point of comparison internally over different time periods as well as with other organizations who similarly tally those same after the fact "lagging" metrics. Given the fact that successfully avoiding personal injury in the interest of self-preservation is an innate instinct in the vast majority of people, it follows that people will normally work more conscientiously in matters involving safety since it is in their own best interests to do so. Incidents involving injury are relatively infrequent and should be relatively easy to count properly.

Because these lagging metrics capture instances where safety efforts have failed, they are really not properly referred to as safety metrics, they are truly nothing more than injury statistics and cumulatively represent the total numbers of circumstances where the organization's safety efforts failed to keep an individual safe from harm. Their greatest weakness is that they are all after the fact measures and are reactive in nature. They don't tell us anything about the nature of an organization's problems or what to do about them. Organizations that focus solely on injury statistics as the only measure of successful safety performance often find their efforts to be shortsighted and this can subject them to losses in areas other than injury; this includes

Applied Operational Excellence for the Oil, Gas, and Process Industries. http://dx.doi.org/10.1016/B978-0-12-802788-2.00008-7
Copyright © 2015 Elsevier Inc. All rights reserved.

fires, business interruptions, and property loss to name but a few. Some will point to the value of learning and applying the lessons that can be gleaned through incident investigation, and while this is a critical function and an important component of every OE program, where is the value in measuring losses if we must wait for a loss to occur before you act? That's reaction—*not* control. The management teams of world class companies work diligently to proactively identify hazards and control the work long before something goes wrong, so as to avoid incidents that result in injury.

Management control involves identifying the work and making the proper efforts required to achieve the desired loss prevention objectives, establishing and maintaining standards for work performance. This involves measuring performance to a specific degree, evaluating performance on a timely basis, communicating it to those accountable, and commending or correcting deficiencies in performance standards. Establishing and tracking the right key performance indicators (KPIs) can help put the proper focus on the activities management expects the workforce to complete. Looking at them in the proper context is always important so there is an understanding of what is getting done and how it is getting done.

Nowadays management has precious little time to read and analyze voluminous data to determine how an organization is performing. Additionally, in this computerized environment of instant communications, where operations seem to move nearly at the speed of thought, there is often a desire for immediate information that will help the line manager review and assess his operations. Structured and administered properly, the use of KPIs can help provide important insights into the current state of operations. Yet, while the KPIs can be useful in many respects, simply reviewing them from a cozy chair behind a desk at headquarters and relying on data viewed on a computer monitor is no substitute for a manager to get out from behind the desk and see for himself what is taking place in the field. The manager that spends time seeing and understanding the operations and engaging the workforce finds tremendous value and has a clear picture of exactly how the operations under his responsibility are running and is then better able to put the data on that computer monitor in the proper context. The importance of a manager or supervisor knowing the operations, the people, the equipment, and the effectiveness of his programs, processes, and procedures cannot be overstated.

Depending upon what is being measured, the information can range from simple and straightforward to very complicated and difficult to compile and analyze. KPIs, are often numbers that are generated from various venues to inform the organizational management of its activities in a "numerical data format." They provide at a glance the relative condition of each discipline the number has been generated for. Businesses can use KPIs to measure progress toward specific health and safety goals or simply to monitor trends associated with corporate and facility activities or special projects. KPIs are used as a means to collect data and communicate trends, which can then be used to indicate where further improvements and resources are required. As is the case with *any* business decision, in order to make good decisions, one must have reliable, accurate and meaningful data. The use of faulty data often complicates decisions and can result in less than optimum results that ultimately increase costs. Measuring the proper indicators and sending the right management signals in support of this are both very important considerations. For example, if the message the management team is signaling to the organization is that they expect their organization to have zero incidents reported, the organization works hard to assure that their results align with this expectation. Unfortunately, human nature being what it is, in an effort to please management and avoid being the one to deliver bad news, incidents and injuries may be hidden or otherwise go unreported. Does the fact that an injury is not reported mean that it did not happen and there are no consequences for the company? Absolutely not—and in fact, failing to properly report and record injury statistics carries with it significant penalties as well as the lost opportunities to properly investigate the incident for the sake of determining basic causes and take the steps necessary to prevent recurrence. Companies recognized as having top tier performances in the petroleum and related industries often make it a point to clearly communicate management's expectations for full and complete reporting of all incidents and injuries and balance that with another very clear signal of their belief and commitment to the principle that zero injuries is attainable. Accordingly, they make that every effort to actively address risks and identify and correct hazards to prevent injuries and losses from occurring.

Selecting good measurements involves considering which characteristics will drive the improvement being sought. Such characteristics typical in OE/SHE performance indicators may include validity, reliability, practicality, and usefulness. Table 8.1 provides a listing of the characteristics of these items and typical concerns with them.

Table 8.1 Characteristics of Good Measurements

Characteristic	Concern	Potential Problem Example
Validity	Does the tool really measure what you think it is measuring? Are you observing the proper indicators? Are indicators being reported accurately?	You are using a satisfaction measurement for a training program when what you are really wanting to learn is the knowledge gain or usefulness of that particular training activity.
Reliability	Does the measurement provide consistent results? Is the number of observations adequate so that the conclusions are accurate? Are all the reporters trained so that their observations will be consistent and reliable?	You might determine that different reporters are not applying the same criteria consistently for the same task or behavior.
Practicality	Is the provided measurement is to use? Is the data easy to obtain, analyze and provide to others?	A complex method of measurement may be established that is too time consuming so is rarely utilized or completed accurately.
Usefulness	Will the data collected be able to be utilized to make meaning improvements? Can others be able to utilize the data to improve?	Data is collected on motor vehicle incidents, but what management needs is data on how effective two particular safety initiatives are at preventing or reducing the motor vehicle incidents.

Most organizations now provide these KPI numbers in a "dashboard" arrangement on an intranet webpage of an organization. Known by various names, including business intelligence front-ends, scorecards, enterprise dashboards, executive dashboards, digital dashboards, KPI reports, dashboards are arranged in numerous configurations. The intuitive nature of business dashboards have become the universal "face of business intelligence." Dashboards allow at-a-glance understanding of the business situation and make for easier decision making. The safety dashboard typically features various safety statistics (e.g., leading and lagging indicators). These are arranged in a dial or graph format, similar to a vehicle dashboard with dials for each indicator, in which real-time statistics (i.e., weekly, monthly reports) are compared to stated safety targets or goals. The dials are color coded, typically with red-orange/yellow-green indicators (loosely correlated to the red/yellow/green indicators of a traffic signal) indicating the relative health of the indicator beside the actual number (See Figure 8.1). Some dashboard designs allow the user to "drill down" to view the underlying supporting data.

Selecting Appropriate Safety KPIs

Which safety KPIs are best for a particular organization depends on several factors:

- Where is the organization today with respect to safety performance?
- Where does the organization want to be in the future for safety performance? Indicators should drive process safety performance improvement and learning.
- Indicators should be easy to implement and easily understood by everyone within the organization.
- The indictors selected should be statistically valid on an industry, company, and site level and be appropriate for the industry, company, or site level for benchmarking purposes.
- Who receives the KPI safety data and what do they do with it?
- How are safety KPIs and their conclusions, communicated to others?

To develop meaningful KPIs, health and safety managers first need to understand the safety risks of their operations, evaluate the systems that are in place to manage risk, and understand the company's business plan and culture. From there line management can decide where they would like the organization to be in the short and long term. It's great to be recognized as one of the industry leaders in the area of safety, but if the organization has a reactive or emerging safety culture, it might want to set a short-term goal of ensuring that it is at the very least in compliance with applicable legal requirements and build from there through continuous improvement. When communicating KPIs, it's also important to understand who the audience will be. For example, establishing a goal of completing all incident investigations within 24 h may help to improve overall safety performance, but it's unlikely to gain favorable attention. One major consulting firm reviewed the most common publicly reported environmental, health, and safety indicators among 44 major companies and found that the companies all reported lagging indicators. Does this mean that these companies are not interested in improving safety performance? No, but it may be an indication that the public doesn't want to know what companies are doing to *improve* performance but is interested in knowing how companies had *performed*. The heavy weighting of companies that track lagging injury statistics may be largely attributed to legal requirements to maintain such records (e.g., US Department of Labor's Occupational Safety and Health Administration (OSHA) requires companies to establish and maintain employee injury/illness records). As for measuring the company's actual performance exclusively through the use of lagging indicators, the reader should bear in mind that injury statistics used in isolation only tell part of the story. Because injury statistics are effectively measures of consequence as opposed to direct measures of control they don't provide a complete picture of the efforts being made to identify hazards and mitigate them; they do however tell you that on a

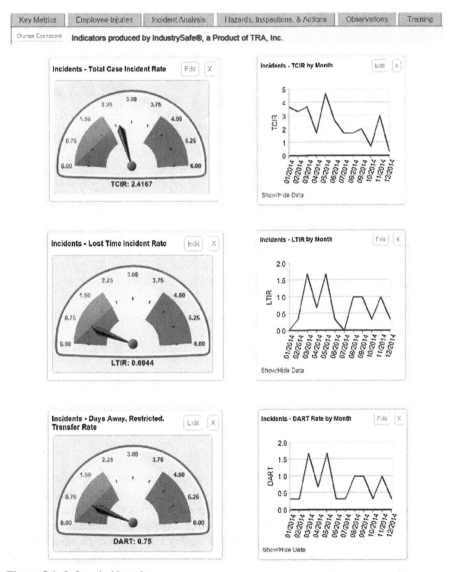

Figure 8.1 Safety dashboards.
Courtesy of IndustrySafe®.

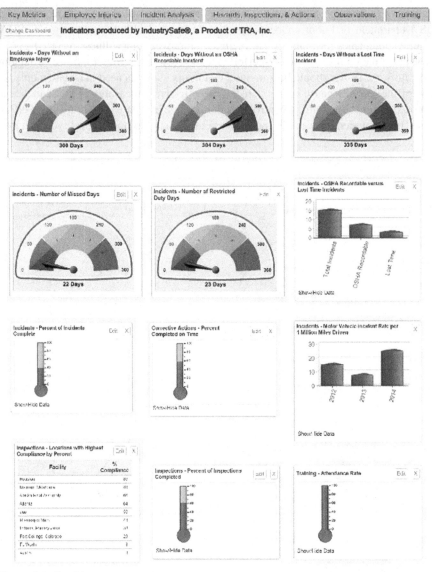

Figure 8.1 Cont'd.

given day, the efforts to prevent such losses associated with that one specific incident were not fully effective. Companies that wait for an injury to occur before they take preventive actions are by definition "reactive" as opposed to "proactive." Such an approach is viewed more as one of "reaction" than "control." This is an important distinction. Operational excellence focuses on management quality and this should connote a more proactive approach to organizing and structuring activities, so as to properly identify and control hazards in a manner that avoids incidents involving injury.

Once the business and its goals have been evaluated, KPIs can be selected. Keep the following points in mind when selecting KPIs:

- You cannot manage what you don't measure.
- Measure the right things for the right reasons.
- Quantity does not equal quality.
- Measure the most important things, not everything.
- Consider piloting metrics before rolling them out company-wide.
- Don't let the cost of measuring exceed the value of the results.
- Consider the KPIs in the proper context.
- Ensure field, line and senior management buy-in.

Leading and Lagging Safety KPIs

KPIs that represent what has already happened are referred to as "lagging" or reactive indicators or measurements. Lagging indicators are commonly used in company communications to provide an overview of performance, such as the tracking of injury statistics, exposure incidents, and regulatory fines, and are commonly considered highly important (see Table 8.3 for additional examples). They are always after-the-fact measurements that focus on past failures and incidents. "Leading indicators" or measurements are more predictive of future performance results. They are viewed as proactive measurements that provide information about the quality, efficiency, or effectiveness of activities, actions, or processes that precede an incident and positively influence safety performance. These might include, among other things (See Table 8.2 for additional examples):

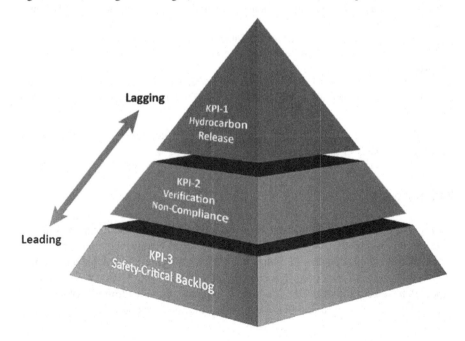

- Number of audits or inspections performed.
- Number and types of findings and observations.
- Timeframe required to resolve and close action items.
- Training completed.
- Near miss incidents reported.
- Timely preventive maintenance tasks performed.
- Safety committee meetings and adequate attendance.
- Crew safety meetings conducted.
- Inspection integrity tests performed on a timely basis.
- Perception surveys.

In most cases, KPIs must be quantifiable and tied to specific targets.

Principles of KPI Measurements

There are several principles when considering the use of KPIs. These include the following:

- The selected KPIs should be aligned with the organizational strategies and priorities. The organization's vision will guide the selected KPIs that will help assessing progress towards achieving the vision.
- KPIs should be measured at all organizational levels.
- KPIs should include both proactive (leading) and reactive (lagging) measurements.
- Measurements should be consistent. The type, recording, period and frequency need to be defined and adhered to.
- Measurement parameters are to be clearly defined without being ambiguous, i.e., they need to be consistent across the organization, e.g., behavior safety observations will be similar from on facility to another.
- A procedure for the method of collecting data needs to be established.
- Measurements should be recorded at time of occurrence. Late recording may allow discrepancies to occur.
- KPIs should be considered as information and not as evidence. This encourages reporting for improvement rather than punishment for employees, otherwise data provided will not be accurate.

All data should be recorded and tracked. This provides for analyses of data (i.e., trending) and basis of improvement.

KPI Targets

Setting targets for lagging indicators is sometimes considered controversial. For instance, it is argued that we should not be setting targets to have incidents and that the target should always be zero. In reality this may not be practically achievable, unless your actual rates are near zero. For example if you have 100 MVAs in one year and expect to have zero the next, this is a highly optimistic though admirable objective. In practical terms it is very unlikely to be achieved unless everyone was to stop driving.

By the same token, setting targets for leading indicators can be problematic as well. For example, if a department sets a target to have completed 50 safety meetings in a year's time and they meet the goal by holding meetings where employees review and discuss issues with little or no bearing on personal or operational safety, will the effort yield the desired impact? And if not, would it be worth measuring and relying on it as a meaningful performance indicator?

Tables 8.2 and 8.3 provide typical examples of leading and lagging safety KPIs used in the process industries.

Table 8.2 Examples of Leading Indicators

Leading Safety KPI
• Average time lag in reporting incidents
• Days without lost time incidents
• Days without OSHA recordable
• Days without incidents
• Employee injuries—top 5 locations
• Employee injury pyramid
• Incident pyramid (e.g. minor to major ratio)
• Incidents by day of week
• Incidents by job title
• Incidents by root cause, cause 1, cause 2, cause 3, and cause 4
• Incidents by type
• Incidents by category
• Incidents by shift
• Incidents by status
• Incidents by worker type
• Incidents by worker age
• Incidents by time with company
• Incidents by time of day
• Injuries by main body Part
• Injuries by detailed body Part
• Injuries by primary cause of injury
• Injuries by detailed cause of injury
• Injuries by nature of injury
• Motor vehicle rate per 1 million miles driven
• OSHA reportable versus lost time
• Total case incident rate
• Number of missed days
• Number of restricted days
• TCIR by month
• Level of near-miss reporting
• Level of deferred maintenance
• Management of change
• Effectiveness of implementation of corrective actions related to incidents

Table 8.3 Examples of Lagging Indicators

Lagging Safety KPIs
• Incidents
• Injuries
• Property damage
• OSHA recordable cases
• Claims paid by claim type
• Claims paid by top 10 locations
• Claims by incident type
• Percent of inspections completed
• Inspection areas with the most deficiencies
• Corrective action status
• Open corrective actions by type
• Corrective actions completed on time
• Most offered training courses
• Top 5 classes by past due
• Training class attendance rate
• Training class pass rate
• Percent of employees with expired training
• Percent of employees no expired training
• Percent observations on schedule
• Percent safe observations
• Percent observations on schedule
• Observation safe/unsafe condition/act
• Observations percent safe by category
• Observations percent safe by subcategory
• Observations top 10 locations
• Hazards by originating type
• Hazards by source
• Hazards by type
• Hazards by evaluation
• Percent of hazards closed
• Percent tasks completed on time
• Liability and litigation costs
• Regulatory citations and penalties

Tiered KPIs

There are three major documents that are relevant to process safety KPIs in the process industries, i.e., API RP 754, *Process Safety Performance Indicators for the Refining and Petrochemical Industries; CCPS, Process Safety Leading and Lagging Safety Metrics; and International Oil and Gas Producers, Recommended Practice on Key Performance Indicators*. All three of these documents utilize a 4 level tiered approach of a safety triangle were serious incidents are at the top (lagging indicators) leading to nonincident events at the bottom (leading indicators), see Figure 8.2. However

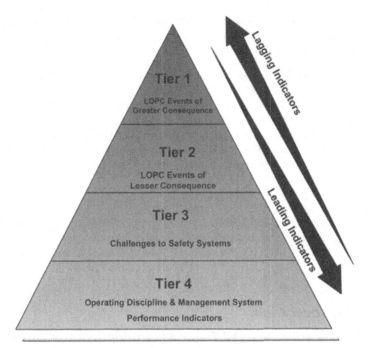

Figure 8.2 Tiered KPIs.

industry professionals have openly questioned whether there is a direct link between occupational (low tier) and process safety (see Limitations of KPIs Measurement Effectiveness). The answer to this is rather lengthy, but suffice it to say that many of the factors that are reflected in adherence to behavioral safety rules through the many layers of rules, standards, procedures, etc. do provide a linkage between the two issues of concern. If the line manager has difficulty in assuring employees meet the established safety expectations in areas often considered to be part of the occupational safety domain (proper use of handrails prevents slips, trips, falls; properly following safety precautions in a material safety data sheet will assure employees are not harmfully exposed to hazardous chemicals), how likely is he going to achieve success in assuring these same employees will properly follow the established procedure for line breaking or isolation of an electrical circuit prior to working on it? The truth is that there is a great deal of overlap between the two and it is not as simple as drawing a line in the sand and proclaiming one as being so distinctively different from the other.

Tier Level Descriptions

Typical Tier Level 1 KPIs

A Tier 1 Process Safety Event (T-1 PSE) is a loss of primary containment (LOPC) with the greatest consequence. It is an unplanned or uncontrolled release of any material, including non-toxic and nonflammable materials (e.g. steam, hot condensate, nitrogen, compressed CO_2 or compressed air), from a process that results in one or more of the consequences listed below:

- An employee, contractor or subcontractor "days away from work" injury and/or fatality;
- A hospital admission and/or fatality of a third-party;
- An officially declared community evacuation or community shelter-in-place;
- A fire or explosion resulting in greater than or equal to typically in highly industrialized countries of approximately $25,000 of direct cost to the Organization;
- A pressure relief device (PRD) discharge to atmosphere whether directly or via a downstream destructive device that results in one or more of the following four consequences:
    - liquid carryover;
    - discharge to a potentially unsafe location;
    - an on-site shelter-in-place;
    - public protective measures (e.g. road closure); and a PRD discharge quantity greater than the acceptable industry practice threshold quantities in any 1-h period;
- A release of material greater than the acceptable industry practice threshold quantities in any 1-h period.

Typical Tier Level 2 KPIs

A Tier 2 Process Safety Event (T-2 PSE) is a LOPC with lesser consequence. A T-2 PSE is an unplanned or uncontrolled release of any material, including nontoxic and nonflammable materials (e.g. steam, hot condensate, nitrogen, compressed CO_2 or compressed air), from a process that results in one or more of the consequences listed below and is not reported in Tier 1:

- An employee, contractor or subcontractor recordable injury;
- A fire or explosion resulting in greater than or equal to typically in highly industrialized countries of approximately $2500 of direct cost to the Organization;
- A pressure relief device (PRD) discharge to atmosphere whether directly or via a downstream destructive device that results in one or more of the following four consequences:
    - liquid carryover;
    - discharge to a potentially unsafe location;
    - an on-site shelter-in-place;
    - public protective measures (e.g. road closure); and a PRD discharge quantity greater than the threshold quantity in any 1-h period; or.
    - a release of material greater than the acceptable industry practice threshold quantities in any 1-h period.

Typical Tier Level 3 KPIs

A Tier 3 PSE typically represents a challenge to the barrier system that had started to progress along the path to harm, but is stopped short of a Tier 1 or Tier 2 LOPC consequence. Indicators at this level provide an additional opportunity to identify and correct weaknesses within the barrier system.

Tier 3 indicators are intended for internal company use and can be used for local (site) public reporting. An organization may use all or some of the example indicators below:

- safe operating limit excursions;
- primary containment inspection or testing results outside acceptable limits;
- demands on safety systems;
- other LOPCs.

or identify others that are meaningful to its operations.

Typical Tier Level 4 KPIs

These indicators typically represent performance of individual components of the barrier system and are comprised of operating discipline and management system performance. Indicators at this level provide an opportunity to identify and correct isolated system weaknesses. Tier 4 indicators are indicative of process safety system weaknesses that may contribute to future Tier 1 or Tier 2 PSEs. Tier 4 indicators can provide opportunities for both learning and systems improvement. Tier 4 indicators are intended for internal organizational use and also for local reporting.

For all of these indicators, nontoxic and nonflammable materials (e.g. steam, hot water, nitrogen, compressed CO_2 or compressed air) have no threshold quantities and are only included in this definition as a result of their potential to result in one of the other consequences.

Industry Practices on KPIs

In 2011, the European Process Safety Centre invited its members to share details on leading indicators that each has introduced for process safety. Seven companies provided this information. A general finding was that no company uses more than six indicators, and one company uses a single indicator as suggested by Table 8.4. The average from this survey was about three (3).

The most commonly found indicators are:

- *Mechanical Integrity*: For example the percentage of inspections completed according to schedule, or the percentage of inspections without nonconformities.
- *Action Item Follow-up*: For example the percentage of actions completed by due date. Differences are apparent in the sources of the action items included in the indicator. Some companies track implementation of corrective actions from incident investigation, others include actions from additional sources (e.g., audits and inspections).
- *Training/Competence*: Often this indicator measures process safety training delivered. As an indicator this might be valid provided the training has process safety specific relevance and individual competence is assessed and documented (by test, demonstration, interview, etc.). Others use it as an indicator for how complete organizational roles in process safety are defined and assigned. Alternatively, one could look at the number of incidents where (lack of) process safety training or competence played a role.

Table 8.4 **Typical Number of Leading Indicators**

Company / Leading Indicator	A	B	C	D	E	F	G	Total
Mechanical Integrity	O	O				O	O	4
Action Item Follow-up	O		O	O		O		4
Management of Change (MOC)	O							1
Training/Competence	O	O		O			O	4
Hazard/Risk Assessment		O	O					2
Overriding/Bypassing		O				O		2
Operating Window Excursions		O				O		2
Activation or Failure of Protective Devices		O			O			2
Number of Leak Boxes and Clamps						O		1
Operating Procedures/Critical Task Execution							O	1
Number of Leading Indicators	4	6	2	2	1	5	3	

The table illustrates other indicators used that proved to be of value for specific companies in their specific situations. When process safety risks are found throughout the operations, leading indicators typically are linked to measuring the functioning of critical safety systems or procedures, such as Management of Change, Risk Assessments, Permit to Work, Mechanical Integrity, etc. It is important to note that quality of indicators, not quantity, is more likely to deliver success.

Major companies that have introduced and are using process safety KPIs have reported the following insights for the benefit of other companies.

- No need to measure everything, start with a pilot. Gather experience in collecting the data, educating, and involving the end-user and in demonstrating the added value in gaining improvement on the selected topics.
- Data from existing systems should be used as much as possible as opposed to the implementation of new and costly systems.
- Whenever several leading indicators are used, it makes sense to aim for a blend of leading indicators comprising both specific operational parameters and functioning of generic safety barriers.
- Crucial for the acceptance may be the choice between a prescriptive approach and one that allows various plants to choose their own relevant leading indicators. This choice need not mean a vast array of indicators that are used across an organization. Experience shows that, when plants are given a choice in selecting indicators, each independently arrives at similar leading indicators.
- It is worth noting that even mature multinational organizations operating with well-established process safety practices and reasonable hazard awareness throughout its workforce will take several years to bring to full operation a working system of leading indicators.

- Leading indicators are often expressed as a percentage or ratio and not an absolute value. They should be expressed positively (100% desired instead of 0%, this is in contrast to lagging indicators). They should promote an informed discussion on where to invest resources (money, effort).
- Use of leading indicators that demonstrates compliance should be avoided, if possible. Any outcome other than 100% compliance is unacceptable and therefore the indicator is not helpful for process safety steering efforts.
- For normalization often the total number of employees and contractor hours in the reporting period is used. It is not directly related to process safety hazards, but is a measure for the scale of operation.
- Leading indicators originate at plant level where the hazards are. They have a greater relevance for operating staff and lend themselves to a greater degree of involvement from the workforce. Only when there are good reasons to compare plants in a company there is a need for shared indicators and normalization (have a so-called common denominator).
- As with any reporting, it will appear that performance is becoming worse before it improves. Prior to reporting there was an impression of the tip of the ice-berg. On reporting the submerged parts become visible. Allow time for sustainable improvement actions.

Managing KPIs for Success

The review of the industry best practices indicates that while setting-up process safety KPIs the following important points need to be considered:

- Process safety KPIs need to be adequately linked to facility's and company's major hazard scenarios
- Identify risks, identify barriers, and develop KPIs
- Leading KPIs be based on measures that prevent major incidents
- At the leadership level a combination of leading and lagging KPIs be reviewed
- KPIs selected for monitoring must have the following characteristics:
  - Reliable
  - Repeatable
  - Consistent
  - Independent of outside influence
  - Relevant
  - Comparable
  - Meaningful
  - Appropriate for the intended audience
  - Timely
  - Easy to use
  - Auditable

For KPIs to be successful there also needs to be a system for tracking, communicating, and improving performance. If data are collected but aren't communicated to the appropriate audience, efforts will not be successful. In an effort to increase accountability the "safety dashboard" is now being commonly utilized to highlight safety performance for leading and lagging indicators in a simple and highly visible manner, in a relatively real-time manner. To date it has been highly successful for increasing accountability and awareness.

Specific Activities for Developing Successful Process Safety KPIs

Establish Organizational Arrangements to Implement Indicators

- Appoint a steward or champion
- Set up an implementation team
- Senior management should be involved

Decide on Scope of the KPIs

- Select the organizational level
- Identify the scope of the KPI measurement system
- Identify incident scenarios—what can go wrong?
- Identify the immediate causes of hazard scenarios
- Review performance and nonconformances

Identify Risk Control Systems and Decide on Outcomes

- What risk control systems are in place?
- Describe the outcome
- Set a lagging indicator
- Follow up deviations from the outcome

Identify Critical Elements of Each Risk Control System

- What are the most important parts of the risk control system?
- Set leading indicators
- Set tolerances
- Follow up deviations from tolerances

Establish Data Collection and Reporting System

- Collect information—ensure information/unit of measurement is available or can be established
- Decide on presentation format

Review

- Review performance of process management system
- Review scope of the KPIs
- Review the tolerances

KPI Evolution for Improvement

KPIs should evolve as the organization changes. SHE managers should be prepared to continuously evaluate their progress in tracking health and safety performance and the benefits of the KPIs. When necessary and appropriate, KPIs should be modified to reflect changing circumstances or to drive further improvement. An organization should always be striving to improve and therefore its KPIs also should advance with improvements. In a positive safety culture, there should be an inverse relationship between proactive and reactive measurements. Therefore if a proactive measurements increase over time, a corresponding reactive measurement will probably decrease.

Limitations of KPIs Measurement Effectiveness

Historical emphasis on lagging indicators has found limitations on the their effectiveness to intervene effectively on future performance, which has led most organizations towards utilizing leading indicators, in addition to lagging indicators. Leading indicators are felt to be more proactive and allow program adjustments before concerns would develop. Additionally, some feel that when the indicator is high, more resources are directed at it and when it's low, less resources are utilized leading to a periodic yo–yo effect in the performance. Additionally some argue that neither leading nor lagging indicators have been proven to be truly effective measures of a program's effectiveness. It is felt they only provide a by-chance intermediate good-sense-of-reason based measures of variance. Much of the published literature does not adequately address the interactive effects between performance variables and the cause-and-effect relationships between leading and lagging indicators.

Additionally with a systems based approach, it is suggested that the driving forces necessary for an effective and successful management system, i.e., management commitment, employee ownership, organizational culture, etc., which need to be measured and tracked have arcane or esoteric characteristics, which leads them to be very difficult to measure. Therefore it may be necessary to supplement these measurements with other non-quantitative techniques. Most utilized, is a survey technique to evaluate if the proper measures are in place. OSHA's Voluntary Safety and Health Program Management Guidelines is an example which offers many programmatic, procedural and behavior examples of observable ways to demonstrate management commitment and employee participation for Safety and Health aspects.

OE Governance and Implementation

Keywords

Administrative processes; Barriers; Challenges; Cost impacts; Documentation; Education; Governance; Implementation; Improvement; Leadership; Learning pyramid; Management; Monitoring; Philosophy; Plan, Do, Act, Check (PDCA) cycle; Planning; Protocols; Quality; Self-improvement; Time impacts.

Cost and Time Impacts of Adopting an OE/SHE Program

The increased emphasis on operational excellence has its roots in the quality movement and the concept introduced with "Total Quality Management," "Continuous Quality Improvement," or any other name given to the quality movement. The common thread running through each of these is not just whether the job was done, but *how* it was accomplished and whether it was effective in meeting the needs, objective and expectations of the organization's stakeholders. All types of industries, including the oil, gas and process industries, have lowered costs and improved the quality of their operations and products by working to meet the needs of the people they serve and providing the proper focus on the way they achieve their results.

Many books have been written describing the philosophy and methods used in the quality movement. The following is a brief overview and relates the basic approach in the common quality management and operational excellence principles by reviewing the work of three leaders who inspired and led the quality movement. These three pioneers all stressed the importance of management awareness and leadership in promoting quality—a concept that is just as critical in the broader pursuit of operational excellence.

W. Edwards Deming

Deming began working in Japan in the early 1950s and is recognized as being instrumental in building the Japanese industry into an economic world power. His strongly humanistic philosophy is based on the idea that problems in a production process are due to flaws in the design of the system, as opposed to being rooted in the motivation or professional commitment of the workforce. Under Deming's approach, quality is maintained and improved when leaders, managers and the workforce understand and commit to constant customer satisfaction through continuous quality improvement.

Applied Operational Excellence for the Oil, Gas, and Process Industries. http://dx.doi.org/10.1016/B978-0-12-802788-2.00009-9
Copyright © 2015 Elsevier Inc. All rights reserved.

Deming and his colleague, Shewhart, promoted the **PDCA** cycle—Plan, Do, Check and Act. World class oil, gas and petro-chemical companies approach operational excellence similarly by applying the continuous improvement concept to the ongoing efforts of improving quality of management and achieving superior operational results.

PLAN to implement a policy to improve quality and/or decrease the cost of providing services. After the plan is developed, we *DO* it by putting the plan into action and then *CHECK* to see if our plan has worked. Finally, we *ACT* either to stabilize the improvement that occurred or to determine what went wrong if the gains we planned for did not materialize. PDCA is a continuous cycle; any improvement realized by carrying out one PDCA cycle will become the baseline for an improvement target on the next PDCA cycle. The process of improvement (PDCA) is never ending, although the dramatic improvements of initial PDCA efforts may be hard to sustain.

The PDCA Cycle

Deming also developed his famous "14 points" to transform management practices. Those points as applied to energy industry operational excellence can be summarized as follows:

1. **Create constancy of purpose**
   An organization's highest priority is to achieve operational excellence across all facets of its operations, cost effectively and in a manner that achieves objectives safely and without harm to the environment. Every organization is responsible to its stakeholders, the community and its own workforce to maintain a high level of excellence and value. Every organization must strive to maximize efficiency and effectiveness through constant improvement.
2. **Adopt the new philosophy**
   Everyone working in the energy industry can find ways to promote quality and efficiency, to improve all aspects of the OE system, and to promote excellence and personal accountability. Pride of workmanship must be emphasized from recruitment to retirement. By their behavior, leaders set the standard for all workers.
3. **Cease dependence on inspection to achieve quality**
   Reliance on routine 100% inspection to improve quality (i.e., a search for errors, problems, or deficiencies) assumes that human performance error or machine failure is highly likely. Instead, there should be a continuous effort to minimize human error and machine failure. As Deming points out, "Inspection (as the sole means) to improve quality is too late!" Lasting quality comes not from inspection, but from systematic improvements in

the underlying processes, programs and procedures. For example, documenting deficiencies in safety record-keeping does not, by itself, generate ideas that would make the task of record-keeping less error-prone. A quality-driven approach might, instead, encourage development of clear and simple record-keeping forms that minimize or eliminate the likelihood of mistakes.

4. **Do not purchase on the basis of price tag alone**
   Purchasers must account for the quality of the item being purchased, as well as the cost. High quality organizations tend to think of their contractors, suppliers, and others as "partners" in their operation. Successful partnerships require clear and specific performance standards and feedback on whether those standards are being met. Contractor and supplier performance can also be improved through an understanding of supplier continuous improvement efforts; longer-term contracts that include explicit milestones for improvement in key features; joint planning for improvement; and joint improvement activities and clear communication of expectations as well as effective coordination and measurement of performance objectives.

5. **Constantly improve the system of production and service**
   OE can be built into all energy industry activities and services and can be assured by continuous examination to identify potential improvements. This requires close cooperation, coordination, and communication between those who provide services and those who consume services. Improved efficiency and service can result from focusing not only on achieving present performance targets, but more importantly, by breaking through existing performance levels to new, higher levels.

The Journey...

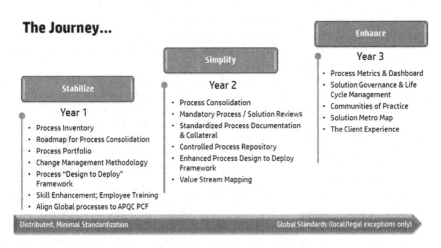

6. **Institute continuous improvement training on the job**
   On-the-job continuous improvement training ensures that every worker has a thorough understanding of: (1) the needs of those who use and/or pay for services; (2) how to meet those needs; and (3) how to improve the system's ability to meet those needs. Incorporating OE into the fabric of each job can speed learning.

7. **Institute effective leadership**
   The job of management is leadership. Effective leaders are thoroughly knowledgeable about the work being done and understand the environment and complexities with which their workers must contend.

Leaders create the opportunity for workers to suggest improvements and act quickly to make needed changes in production process. Leaders are concerned with success as much as with failure and focus not only on understanding "substandard," but also "super-standard" performance. The effective leader also creates opportunities for below- and above-average performers to interact and identify opportunities for improvement.

8. Drive out fear

The Japanese have a saying: "Every defect is a treasure," meaning that errors and failures are opportunities for improvement. Errors or problems can help identify more fundamental or systemic root causes and ways to improve the system.

Yet, fear of identifying problems or needed changes can kill continuous improvement programs! Also, some may feel that the idea of making improvements is an admission that the current way of doing things is flawed or that those responsible are poor performers.

Improved performance cannot occur unless workers feel comfortable that they can speak truthfully and are confident that their suggestions will be taken seriously. Managers and workforce members must assume that everyone in the organization is on board with making the OE system work and is interested in doing his best!

9. Break down barriers between departments

Barriers between organizations or between departments within one organization are obstacles to effective improvements. Inter-departmental or intra-organizational friction or lack of cooperation result in waste, errors, delay, and unnecessary duplication of effort. A continuous and lasting continuous improvement program requires teamwork that crosses traditional organizational lines. Continuous improvement requires that all workforce members, business lines, administrative areas, departments, and units share a unified purpose, direction, and commitment to improve the organization. Intra-organizational pathways are developed and cultivated as mechanisms by which to improve performance.

10. Eliminate slogans, exhortations, and targets for the workforce for zero defects and new levels of productivity

The problem with such exhortations is that they put the burden for quality on worker performance instead of poor system design. Continuous improvement requires that the organization focus on improving its work processes. In so doing, quality, efficiency and effectiveness will increase, productivity will rise, and waste will diminish.

11. Eliminate management by numbers and objective. Substitute leadership!

For Deming, work production standards and rates, tied to incentive pay, are inappropriate because they burn out the workforce in the long run. Alternatively, a team effort should be marshaled to increase quality, which will lead to increased profits/savings that can then be translated to, for example, higher salaries or better benefits. Improvement efforts should emphasize improving processes; the outcome numbers will change as a consequence.

12. Remove barriers to pride of workmanship

The workforce is the most important component of an organization's OE system. Companies in the oil, gas and chemical industries cannot function properly without workers who are proud of their work and who feel respected as individuals and professionals. Managers can help workers be successful by making sure that job responsibilities and performance standards are clearly understood; building strong relationships between management and the workforce; and providing workers with the best tools, instruments, supplies, and information possible.

13. Institute a vigorous program of education and self-improvement

Energy industry workers can improve their lives through education and ever-broadening career and life opportunities. The energy industry needs not just good people; it needs people who are growing through education and life experiences. Management, as well as members of the workforce, must continue to experience new learning and growth.

14. Put everybody to work to accomplish the transformation
At the heart of continuous improvement is an organization-wide focus on meeting the organization's objectives and expectations efficiently, effectively and in a repeatable and reliable manner. Effective OE management programs go beyond emphasizing one or two efforts or areas to improve performance. Every activity, every process and every job in OEMS can be improved. Everyone within the organization can be given an opportunity to understand the continuous improvement program and their individual role within that effort. Improvement teams that include broad representation throughout the organization can help ensure success of initial efforts and create opportunities for cross-disciplinary dialogue and information exchange.

Philip Crosby. Crosby coined the phrase "quality is free," meaning that the absence or lack of quality is costly to an organization, e.g., in money spent on doing things wrong, over, or inefficiently. Conversely, spending money to improve quality, e.g., to reduce waste or improve efficiency, saves money in the long run.

According to Crosby, ensuring quality should occur primarily at the design phase. Rather than spending time and money on finding and fixing mistakes and errors, Crosby advocates organizational changes to encourage doing a job right the first time. Crosby challenges organizations to think of how processes can be designed or re-designed to reduce errors and defects to reach a goal of "zero defects."

Crosby believes managers' policies and actions indicate their commitment to quality. He also advocates a step-by-step approach for educating the entire workforce about quality principles, extensive measurement to document system failures, and formal programs to redesign faulty production processes.

Joseph Juran. Juran's approach is based on the idea that the quality improvement program must reflect the strong inter-dependency that exists among all of the operations within an organization's production processes.

According to Juran, *Quality Planning* is the process of understanding what the customer needs and designing all aspects of a system that is able to meet those needs reliably. Designing an OE system to do anything less is wasteful because it does not meet stakeholder needs or expectations. Once the system is put into operation, *Quality Control* is used to constantly monitor performance for compliance with the original design standards. If performance falls short of the standard, plans are put into action to deal quickly with the problem. Quality control puts the system back into a state of "control," i.e., the way it was designed to operate. *Quality Improvement* occurs when new, previously unattained, levels of performance—breakthrough performance is achieved!

Juran also proposed the idea of the "Vital Few and the Useful Many" that helps prioritize which continuous improvement projects should be undertaken and in which order. In any organization, there will be a lengthy list of possible ideas for improvement. Since the resources to actually implement new ideas is limited, however, leaders must choose those **vital few projects** that will have the greatest impact on improving ability to meet customer needs. The criteria for selecting continuous improvement projects includes potential impact on meeting customer needs, cutting waste, or marshaling the necessary resources required by the project or task.

Juran also developed the idea of instituting a leadership group or "Quality Council," consisting of the organization's senior executive staff. The Quality Council is typically charged with the responsibility for designing the overall strategy for quality planning, control and improvement. Senior leadership involvement is a must, since continuous improvement activities are as important as other management tasks (e.g., budgeting, human resource management, purchasing, and training), and leaders can integrate continuous improvement into every aspect of energy industry operations.

The principles and methods advocated by Deming, Juran and Crosby provide a basic foundation for OE and most of today's associated continuous improvement efforts. Reading about the work of others provides a start, but in the long run, it will be unwavering leadership that will provide the most significant ingredient for success. That leadership can be achieved through a personal and professional commitment to learn and apply these principles.

Governance

OE Governance describes the entity that will oversee OE systems design, coordinate external assessments and monitor implementation and support organization. There are two levels of governance; local governance at the business line, administrative or department level and corporate OE governance at the executive management level.

Good corporate governance also means having organizational structures and management processes in place to make sure that the company's decisions and actions are in the best interests of their stockholders. It should be transparent and responsive to stockholders. This is achieved transparently and responsively through reports, press releases, the organization's Website and meetings to discuss governance, financial, environmental, social and policy issues. Stockholders can also direct inquiries to the Board of Directors and submit proposals for inclusion in the proxy statement.

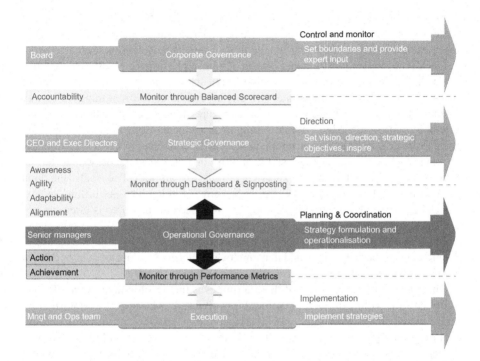

Implementation

An OE program can be organized like any other management program.

The main steps include the following:

1. Determine the objectives of the program
2. Create the Organization
3. Develop a Plan
4. Implement the Plan
5. Audit and Improve

Management consultants indicate that more than 70% of all major transformation efforts fail. Why? Because organizations do not take a consistent, holistic approach to changing themselves, nor do they engage their workforces effectively. They also found that companies that became particularly adept at continuous change have three characteristics in common.

These organizations:

- Closely followed the process
- Allowed the process to flex itself to their specific contexts
- Measured their efforts in terms of results

These organizations mastered a capability that greatly enhanced their likelihood of success in the present and the future. It also helped them survive: organizations that do not or cannot evolve in the permanent context of rapid change will not last.

One of the single most influence effects of human performance is attitude. Attitude is a reflection of the employee's mindset, motivation, point of view, cultural influences, and it generally demonstrates the way an individual looks at things. The way we look at things is partly responsible for the nature of our behavior and performance. A poor attitude may lead to errors resulting in an incident. Many organizations today strive to create a learning organization, which can be defined as one that is continually learning new KSAs (knowledge, skills, abilities and attitudes). Considerable formal effort is applied to increasing our knowledge, skills and abilities. The results can often be measured with a picture of "where are we now," through key performing indicators (KPIs). In the KSAs group, the one not easily seen or measured is often the one that enables or impairs the learning organization. This is attitude. Why are workers' attitudes so important? Because they're the route to safe behavior within the organization. Recent studies have indicated that employee "success" factors are related 85% to behavior features and only 15% to skills.

The best treatment for most of these attitudes is the development of an effective safety culture within the organization. A positive safety culture is exemplified by senior management with employee involvement that demonstrates the mutual benefits of an incident free environment for both the organization and the individual. Sometimes supervisors find they can't always directly influence workers' behavior. Rules may not work; training may not work. But attitudes typically drive behavior. People learn by watching others. They pay attention to what others do and what they say are teaching tools. They learn what is "acceptable behavior" and what is not through the actions of management extending accountability and through validation by their peers. Workers' attitudes reflect their evaluation of what they have learned. And rather than saying that all workers are rational and logical and act accordingly at all times, because we understand humans are by their nature complex beings, it is sufficient to say that their feelings and emotions are also variables that factor into attitudes at any given time. These can directly or indirectly impact their behaviors. Individuals that harbor resentment towards their supervisors because they perceive favoritism to another, someone that feels slighted in their last performance evaluation, or the employee that just had a marital tiff with their spouse before they left for work are all examples of those who may allow their feelings and emotions influence their attitude on the job. In some cases it may only result in a slight distraction, or the buildup of resentment, animosity and/or anger, but even so, such variables may not be outwardly evident and be difficult for the supervisor to discern and address. Certainly, knowing the people in his organization makes this task at least somewhat easier for the supervisor and demonstrating genuine care and concern for the safety and well-being of the individual as well as treating him fairly with respect goes a long way towards creating a positive upbeat and safe work environment.

Employees can help change other's attitudes by their own beliefs, behaviors and the attitude they exhibit towards those beliefs. If they believe that incidents in the workplace can be prevented, others in the workplace attitudes (and behaviors) will reflect that belief. Their behaviors and attitude in turn will affect what other workers believe and, ultimately, how they behave. What I do, what I say and how I say it can change lives and prevent an incident occurring. Organizations that can effect this change in attitude are true learning organizations.

Learning Pyramid

A learning pyramid (See Figure 9.1) graphically represents the various methods an individual can undertake to learn information at various levels of retention. It illustrates the various levels of retention or recall for an individual by the various approaches that can be utilized. The first four levels, i.e., lecture, reading, audio visual, and demonstration, are passive learning methods. While at the lower three levels, are located discussion groups, practices by doing and teach others are active or participative learning activities. The research that undertook this analysis indicates that active participation helps to greatly retain the information to be learned. The difference in retention between active and passive means is thought to be attributed to the long periods of reflection and deep cogitative processing.

This is why much emphasis is placed in undertaking "walk the talk" approach in proactive learning organizations.

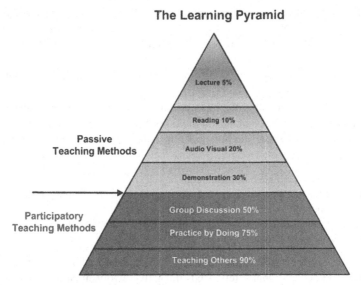

The Learning Pyramid

Figure 9.1 The learning pyramid.

Safety Culture

The term safety culture is often over used and frequently not well understood. To some people safety culture means the company has rules that have to be followed and how that is perceived is largely a result of the leadership in place. Safety culture is defined by what an organization does and the actions it takes routinely and reliably, more so than just the words used to portray the desired state. Words without actions ring hollow and so it is with a company's core safety and the overall operational excellence culture. Here are some diagnostic questions that may help determine where your company safety culture currently stands:

- Does a manager and his supervisory staff provide the necessary leadership that makes it clear that safety is a value and the safety and ensuring the well-being of every single worker on the site of critical importance and something that is in every single person's own self-interest?
- Does everyone perceive that doing things the right way and not taking shortcuts is the right way to approach every task that must be completed?
- Do workers trust the management team to make good decisions that are in the best interest and support the safety and well-being of every member of the workforce?
- Do all workers understand the expectation for approaching their work with a planning mindset, the importance of figuring out what can go wrong before starting and taking the proper precautions to protect themselves and others before proceeding?

These and other questions should begin to indicate that safety leadership is foundational to the operational excellence journey. It has been said that a company's safety culture is not what they say they will do (as in their written procedures, rules, and standards) but what everyone does in actual fact, when no one is looking over their shoulders, how they approach each and every task, day in and day out, 24/7.

Looking a little bit closer at how the OE safety culture fits into corporate governance, we would like reflect on and summarize portions of the good work presented by the Organization for Economic Co-operation and Development (OECD) in 2012. The work of the OECD related to corporate governance for process safety was carried out by the Working Group on Chemical Accidents (WGCA) and it focused on works in three primary areas: response, analyzing issues of mutual concern and making recommendations on best practices.

Leaders need to understand the risks posed by their organization's activities, and balance major accident risks alongside the other business threats. Even though major accidents occur infrequently, the potential consequences are so high that leaders need to recognize:

- Major accidents as credible business risks;
- The integrated nature of many major hazard businesses—including the potential for supply chain disruption;
- Management of process safety risks should have equal focus with other business processes including financial governance, markets, and investment decisions, etc.

Comprehensive operational excellence incorporates and addresses aspects of both operational and occupational safety needs the active involvement of senior leaders. It is very important that they are highly visible within their organization, because of the major influence they have on the overall safety and organizational culture.

To maintain the right focus on running the business properly and preventing major accidents, leaders also need to recognize the full extent of the impact of these incidents and the potentially devastating consequences for a business, including

- Harm to people, including loss of life and serious injury;
- Environmental damage—or example air, water and land contamination;
- The damage to business efficiency from disruption of production, and loss of customers or suppliers/contractors;
- The potentially huge costs—both direct (for example asset replacement or repair costs, legal fees, and fines) and indirect (for example increased insurance premiums, loss of shareholder confidence resulting in falling share value, loss of public confidence in the company to supply products and switching to other brands, etc.);
- Negative effects on the local economy;
- Long-term damage to an organization's reputation, from adverse publicity, legal action and harm;
- Potentially increased governmental regulation and oversight;
- To the company "brand," and
- The discontinuation of the company as a viable, ongoing entity in light of the above.

However, good corporate governance of operational safety is not just about avoiding potential negative effects. There are numerous commercial reasons why effective safety management makes good business sense.

Some of the benefits of well managed assets and processes include:

- Less downtime, and higher plant availability;
- Maintenance budgets that are easier to forecast;
- Plants and equipment which have longer life spans;
- Improved efficiency and flexibility;
- Enhanced employee, stakeholder and regulator relationships, and
- Access to capital and insurance at more attractive rates.

These factors allow production scheduling to run more smoothly and help create a better, more productive business, with a less stressful working environment for managers and employees alike.

Key Self-Check Questions for the Senior Manager

- Do you know what the major accident risks are for your organization?
- Do you know what your main vulnerabilities are?
- What are you doing about them?
- How concerned are you about the level of risk?
- How confident are you that all the safety systems are performing as they should?
- Do you seek out the "bad news," as well as the good?
- If there is an incident, who do you blame? Others or yourself?
- Are you doing all you can to prevent a major accident?

Essential Elements of Corporate Governance for Safety

Strong leadership is vital, because it is central to the culture of an organization, and it is the prevailing culture which reflects and strongly influences employee behavior and achieves the target safety performance. Functional safety tasks (including those relating to process or operational safety) may be delegated, but responsibility and accountability will always remain with the senior leaders, so it is essential that they promote an operating environment which encourages safe behavior.

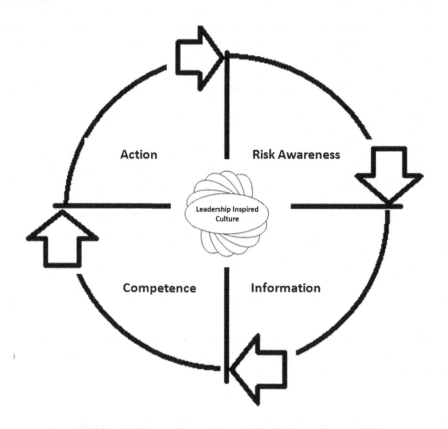

OE Leadership and Culture: The CEO and every member of his management chain of command create an open environment outlined in the following table:

Leadership and Culture: Leadership creates an open environment where they:

Keep all aspects of safety (both operational and occupational aspects) on their agenda, prioritize it strongly and remain **mindful of what can go wrong.**

Encourage people to raise operational safety concerns, or bad news to be addressed.

Take every opportunity **to be role models,** promoting and discussing safety.

Delegate appropriate process safety duties to competent personnel while establishing and maintaining overall **responsibility and accountability.**

Be visibly present in their businesses and at their sites, asking appropriate questions and constantly challenging the organization to find areas of weakness and opportunities for continuous improvement.

Promote a "safety culture"—where **each task is completed the right way—every time,** such that is known and accepted throughout the enterprise.

Demonstrate the need **and commitment to identifying and correcting hazards, promptly taking action to resolve deficiencies** and providing a focus on continuous improvement.

Leadership Inspired Culture

Risk Awareness: Leadership broadly understands the vulnerabilities and risks and they:

Know the importance of operational safety throughout the asset's life cycle—throughout the design, construction, operation, and maintenance phases of their manufacturing facilities, including decommissioning at those locations.

Understand the critical and different layers of protection that are in place between a hazard and unsafe condition that can lead to an incident **continually** seek to strengthen those layers.

Ensure appropriate and **effective management system for analyzing, prioritizing and managing risk,** including strong **management of change processes** for equipment, people, technology and facilities.

Personally involve themselves in risk assessing proposed budget reductions for safety impacts and provide incentive schemes which **don't encourage production at the expense of operational safety risks.**

Take responsibility for emergency planning for the range of consequences from an operational safety incident including the credible worst case scenario.

Know the hazards and risks at installations where there are hazardous substances.

Risk Awareness

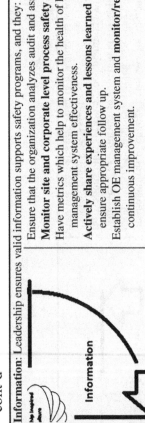

Information	**Information:** Leadership ensures valid information supports safety programs, and they: Ensure that the organization analyzes audit and assessment results. **Monitor site and corporate level process safety key performance indicators and near misses.** Have metrics which help to monitor the health of key safety processes, positive safety culture (and problem areas) and OE management system effectiveness. **Actively share experiences and lessons learned** within their own organization and within other energy industry sectors and ensure appropriate follow up. Establish OE management system and **monitor/review its implementation and on-going effectiveness.** Aggressively pursue continuous improvement.
Competence	**Competence:** Leadership assures workforce competence to properly address hazards, they: Understand which questions to **ask their people** and know which follow up actions are necessary. **Ensure there are competent management, engineering, and operational personnel at all levels.** Ensure continual development of operational safety expertise and learning from new standards, regulation and best practices. Ensure continual development of process safety expertise and learning from new regulation and guidance. **Provide resource and time** for hazard and risk analysis, effective training and comprehensive scenario planning for potential incidents. Defer to the expertise of personnel, and do not dismiss expert opinions. They provide a process or system to ensure company leaders get expert operational safety input as a critical part of the decision making process for commercial projects or activities. Ensure that the organization monitors and **reviews the safety competency of contractors** and third parties. Are capable of openly communicating critical aspects of operational safety with all internal and external audiences.
Action	**Action:** Leadership engages in articulating and driving active monitoring and plans, they: Ensure **practices are consistent with corporate process safety policies.** Hazard mitigation measures should be incorporated at the earliest possible conceptual and engineering design stage of an installation to enhance the intrinsic (inherent) safety of the installation wherever practicable. Incorporate operational safety considerations into major capital investments, long range planning and integration of mergers or acquisitions. Ensure safety risk mitigation plans and **emergency response plans are developed and maintained for all sites** within their business and at an organization-wide level, with appropriate levels of competent resources available to execute the plans. Ensure implementation of operational safety risk mitigation plans and reviews of progress against the plans at site and corporate level. **Monitor implementation of proper corrective actions** and ensure they are promptly completed following audits and after thorough root cause investigations of all incidents or potentially high consequence near misses.

Suggestions for Implementing Change in an Organization

As a business creates a positive safety culture, there are always instances where doing the right thing will meet with some initial resistance. This is a natural human reaction to any type of change. Leadership can start with some simple tactics to influence to achieve a healthy change in the organization which include:

- Start with changes that you know will be accepted and successful. Then follow on and build on these successes.
- Focus on the people who are the believers in the changes, not on the skeptics.
- Let small achievements give you momentum and then work on the more difficult and tougher changes.
- When making changes do it in small increments so that eventually a large change will occur.
- Obtain the support of key individuals and leaders, not just management.
- Be a leader yourself. Create relationships and trust by helping others understand and accept the changes.
- Ensure your own actions and activities are consistent with the requirements of the changes.
- Think positively. Change is always slow, but does happen. It is part of the continuous improvement process.

The following symptoms are common with rolling out a new management system. It is important to anticipate, recognize and understand the symptoms and the problems to be able to correct them.

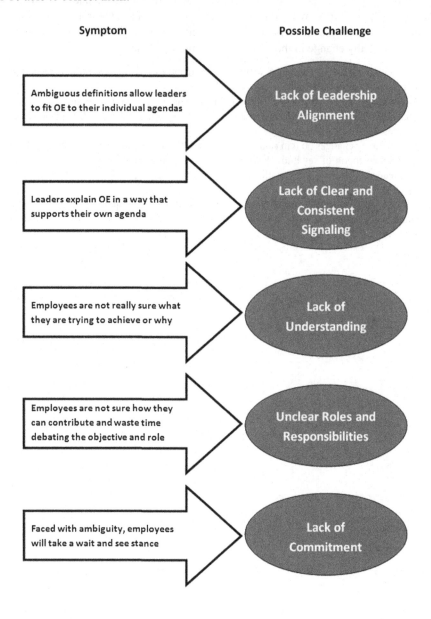

The following checklist (Table 9.1) provides examples of priming behaviors for the line manager to think through and use to develop his own plan for demonstrating the proper OE leadership behaviors.

Table 9.1 Operational Excellence Leadership Priming Behaviors

Visible Care & Commitment	Think About What to Say and Do	Establish an Observation Frequency and Engage
Demonstrate personal care and concern for the health and safety of every individual and for the protection of the environment.		
Model OE behaviors that integrate safety values, principles and beliefs into daily work and hold others accountable to do the same. Always "walk the talk."		
Ensure that the entire workforce understands, believes and uniformly demonstrates through proactive measures that every job and every task can be completed every day without injury or incident; *do it safely or not at all.*		
Invest personal time and attention to ensure programs, processes and procedures needed for managing safe, reliable, and efficient operations are working effectively.		
Extending Accountability for OE Performance		
Establish a vision and set clear world-class OE objectives for the organization.		
Simplify messages and integrate and widely communicate performance objectives, metrics and targets to every level for: • OEMS implementation • Achievement of OE expectations through process development and use performance expectations • Improvement of OE behaviors		

Continued

Table 9.1 **Operational Excellence Leadership Priming Behaviors—cont'd**

Visible Care & Commitment	Think About What to Say and Do	Establish an Observation Frequency and Engage
Ensure work is prioritized, resources are available, and that roles, responsibilities and accountabilities are fully aligned throughout the organization		
Assign responsibility and direct and monitor the implementation of OE plans to their successful conclusion.		
Provide appropriate positive and negative consequences for: • OEMS implementation • Achievement of OE expectations through process development and use achievement of results • Improvement of OE behaviors		
Measure and Verify		
Visit operations frequently, engage people and take time to reinforce OE performance and test OEMS effectiveness. Review the effectiveness of OE plans, processes, OEMS implementation, and revise as necessary. Actively follow through on incident investigations to ensure that root causes are determined and mitigating actions are carried out.		
Know operations well enough to identify workplace hazards and verify that critical OE processes and practices are working as intended.		

Performing an OE/SHE Gap Analysis

10

Keywords

Barriers; Challenges; Continuous improvement; Deficiencies; Evaluation; Expectation; Gap analysis; Management of change; Measurement; Monitoring; OE/SHE adherence; Opportunities; PDCA cycle; Performance improvement cycle; Performance improvement targets; Quality; Review; Strategic planning; Strengths; SWOT analysis; Threats; Weakness.

Gap analysis is a technique that organizations use to determine the steps needed in order to move from its current state to a specified future state. While the number of steps may vary from that shown below, it serves to graphically illustrate the stair step nature of continuous improvement and the use of gap analysis.

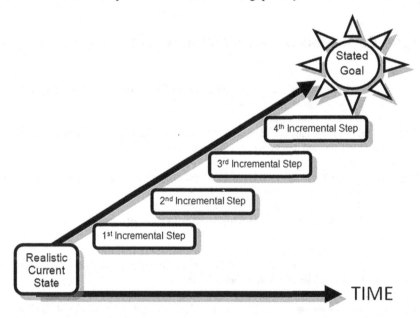

Organizations use gap analysis to candidly assess where they are at and what sort of incremental efforts are necessary to get them to their stated goal or objective. It can help identify the barriers, challenges and/or deficiencies that exist in between the current state of affairs and the determined future state. Some companies will define the future state by examining and projecting what is possible through relatively small incremental year-on-year improvements (e.g., 3–5%) performance improvement targets, gains possible through bottom-up alignment (aggregating unit level targets to

Applied Operational Excellence for the Oil, Gas, and Process Industries. http://dx.doi.org/10.1016/B978-0-12-802788-2.00010-5
Copyright © 2015 Elsevier Inc. All rights reserved.

division and department level) as well as possible improvements possible through internal and external benchmarking (identifying, adapting best practices for process performance improvements).

Who conducts the gap analysis? Typically these will be completed at the department level by local teams consisting of management and/or individual process owners. They are typically done on an annual basis as part of a self-assessment exercise but can be set for completion on various frequencies as deemed appropriate. The gap analysis should be completed by a team with technical knowledge, credibility with peers, high energy, enthusiasm, and sound logical reasoning ability. The gap analysis is most often done in conjunction with the Plan, Do, Check and Act (PDCA) cycle. As mentioned previously, the PDCA cycle involves management activities throughout the Plan, Do, Check and Act phases and these are all integral to the continuous improvement process. Continuous improvement efforts are pursued repeatedly and with regularity to help maintain focus on progress towards specific goals. Continuous improvement never ends and this is one of the keys to Operational Excellence, there must be relentless diligent pursuit of continuous improvement for the organization to achieve excellence and recognize some of the competitive advantages that accompany it.

Typical Performance Improvement Cycle

1. *Intent*
   Leaders provide the vision and set the expectations for operating performance through a local operating policy and consistent actions.
2. *Risk Assessment and Prioritization*
   Risks and performance gaps are identified and opportunities for improvement are prioritized with applicable legal and regulatory requirements met.
3. *Planning and Controls*
   Plans establish clarity about intended activity and controls confirm that objectives are sustainably achieved.
4. *Implementation and Operation*
   Activities are carried out consistent with the plan to meet commitments as well as legal requirements.
5. *Measurement, Evaluation and Corrective Action*
   Monitoring and measurement are carried out to determine if applicable requirements and plan targets are being met and controls are effective.
6. *Management Review and Improvement*
   Management verify the statement of intent is being met, and review the local OMS implementing any identified changes.

Evaluation

An ongoing evaluation is essential to ensure that the expectation in the framework are being meet. OE employs internal and external assessment processes to gage the degree to which expectations are being achieved. Such evaluations provide evaluations to determine areas for further improvements both in performance and supportive management systems. The assessment process focuses on the evaluation of management systems and their individual components. Two major system aspects are evaluated.

System Status

- Extent to which the five characteristics of an operational excellence system are incorporated in the system design and properly documented.
- Extent of development, which includes communication, training and measurement establishment.

System Effectiveness

- Extent of conformance to system requirements and documentation.
- Quality of system execution.
- Determination of how well the system is operating and if the stated objectives are being achieved.

Audits are undertaken to ensure OE programs and processes are properly implemented and maintained and to provide information to management for improvements. The OE audit should include the objective and criteria to be used, activities, and areas to be covered, frequency of audits, composition and leadership of the audit team, how the audits will be conducted and method and format of audit reports (see Figure 10.1 and Table 10.1).

The OE audit report should also indicate the method to respond to any findings and how these findings will be recorded and tracked until resolved and closed.

A graphically representation is usually utilized as it is easy to understand and display the identified gaps in an organization. Figure 10.2 provides a useful example showing a graphic portrayal of a gap analysis.

Gap analysis is a strategic planning tool to help you understand where you are, where you want to be, and how you're going to get there. Gap analysis relies upon taking a critical look at what the current standing or situation is for a person to a company. In order to make any improvements in a company, the first step is to understand where you are and where you want to be. You can understand where you want to be by looking at the company's mission statement, strategic objectives, and improvement goals. The first step, then, in performing a gap analysis is to define where you want to go using terms as specific as possible.

Management of Change (MOC)		
Is MOC utilized at this facility?	Yes	No
Facility	Area	
Documents Reviewed	Titles	Date
Employees Interviewed	Name/Title	Date
Field Observations		
Loss Prevention Reviews MOCs with hazard assessments?		
Notes		
Reference Material		

Figure 10.1 Management of change audit form.

Table 10.1 **Safety Department Performance Expectations**

Safety Department Performance Expectation	Excellent	Above Average	Average	Below Average	Poor
Assistance/support with safety programs and processes—development and implementation.					
Safety standards, policies and instructions, construction safety requirements and related safety requirements (i.e., useful, practical and cost-effective).					
Participation in project reviews, i.e., project proposals, detailed designs, PSSRs, MOCs, commissioning, etc.					
Support to help proponents address contractor safety.					
Field support (audits, inspections, drills, consultations, etc.).					
Sharing of incident findings and lessons learned.					
Engagement/support with members of proponent management.					
Safety publications (i.e., topics, depth of information, frequency, etc.).					
Quantitative risk and consequence modeling services.					
Safety campaigns (e.g., fall protection, work permits).					

After putting together the "where we want to be" against the "where we are," you can start to see the different steps that will need to be taken in order to bring about the desired effects. If you see areas where your company is particularly strong, ask yourself how processes there can be implemented in the areas that need improvement.

This is the simplest gap analysis method.

Step 1: Decide the topic you're going to do the Gap analysis on. This is the challenge you're trying to tackle.

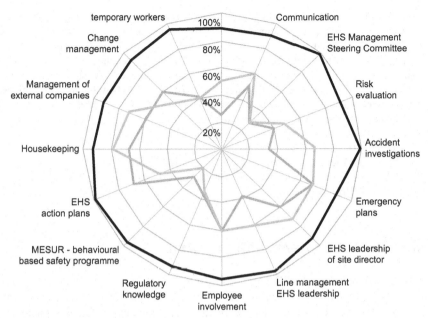

Figure 10.2 Graphical portrayal of a gap analysis.

The Gap analysis can (and should) be done for virtually all processes, programs, and activities that comprise the OEMS. For the sake of simplicity we present several sample topics to serve as examples:

- Revenue
- Profit
- Market Share
- Product Functionality/Features

Step 2: Identify where you are right now based on metrics or attributes.
Examples:

- Revenue—We're at $25 million in annual sales right now
- Profit—We're at $5 million in annual profit right now
- Market Share—We have a market share of 10% right now
- Product Functionality/Features—Our product has wide application in the transportation field.

Step 3: Identify where you'd like to be over a specific time frame.
Examples:

- Revenue—We would like the organization's revenue to grow to $50 million in annual sales by 2025.
- Profit—We'd like profits to grow to $15 million per year by 2025
- Market Share—We'd like to own 20% of a particular market by 2025
- Product Functionality/Features—We'd like our product to have industry leading features by 2025

Step 4: Identify the gap between where you are and where you want to be.

- Revenue—The gap is $15 million per year in annual sales by 2025
- Profit—The gap is $5 million in annual profit by 2025
- Market share—The gap is 20% market share by 2025

- Product Functionality/Features (let's use the website as an example)—The gap is that you'd like to have the following features by 2025: a blog, a sign-up form to let visitors follow your business on Facebook and Twitter, and a way for customers to buy products directly.

Step 5: Determine how the Gap should be filled. This is the step that requires forethought and planning to make the incremental changes (at times referred to as step change) to continuously drive performance to the desired levels. The plans can be very simple or complex but in essence they will typically involve some combination of the "6 M's":

- Manpower—The people resources you need.
- Methods—The processes you need.
- Metrics—The measurements you need.
- Machines—The automation or technology you need.
- Materials—The material items (such as physical goods or marketing collateral) you need.
- Minutes—The time you need.

Prioritize Action Items and Initiatives

For the action items use the PDCA as your prioritization criteria. Action items developed from the "Plan" gap analysis should have the highest priority since they are essential before moving to the "Do" stage. Similarly the "Do" related actions items should precede the "Check" items and so on.

For the initiatives, use the "Impact versus Easiness" matrix to prioritize them. This method uses a set of qualitative and quantitative measures to assign a priority score to each initiative.

To demonstrate, in Figure 10.3 each dot represents an initiative. The figure has four quadrants. Initiatives that fall in the upper right quadrant will have the highest priority and should be part of the first implementation wave. Waves 2, 3 and 4 will follow respectively.

Following are suggested guidelines to determine initiative impact and ease of implementation (see Table 10.2).

Defining Potential Impact: the potential impact of an initiative should be analyzed using selected criteria such as the ones shown in table below. Selection criteria can include initiative impact on department profitability, satisfying operational urgency or other qualitative benefits such as customer satisfaction or employee morale.

Analyzing Ease of Implementation: Ease of initiative implementation could be analyzed using criteria in the table below. Selection criteria can include the cost of initiative (the lower the better), resources requirement, and time span of initiative (Table 10.3).

SWOT Analysis

In completing the assessment there are other tools that can help guide the analysis that will allow for plan development. One such tool is called an SWOT analysis. Quite simply it involves listing out Strengths, Weakness, Opportunities and Threats related to filling your Gap.

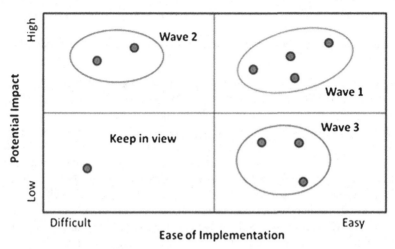

Figure 10.3 Goal prioritization.

Table 10.2 **Potential Impact Evaluation Factors**

	Low	Medium	High
Revenue generation/cost saving (annual revenue/ saving in $M)	<100	100 to 250	>250
Impact on HSE	Minor enhancement	Moderate enhancement	Major enhancement
Qualitative improvement (e.g., employee/customer satisfaction, knowledge sharing, etc.)	Low impact	Medium impact	High impact

Table 10.3 **Implementation Evaluation Factors**

	Difficult	Moderate	Easy
Cost (NDE of implementation in $M)	>250	100 to 250	>100
Resources (people required to implement initiative)	>50	10 to 50	<10
Timeframe (time until benefits are realized)	>6 months	3–6 months	<3 months

SWOT is a strategic planning tool. The acronym SWOT stands for:

- *Strengths*—These are attributes that you or your company possess that would be helpful in achieving the objective.
- *Weaknesses*—These are attributes that you or your business have that are harmful to achieving the objective.

- *Opportunities*—These are the external circumstances that are helpful to you achieving the objective.
- *Threats*—These are external circumstances that could damage the performance of your objective.

The SWOT analysis template is normally presented as a grid, comprising four sections, one for each of the SWOT headings: Strengths, Weaknesses, Opportunities, and Threats. The SWOT template below includes sample questions, whose answers are inserted into the relevant section of the SWOT grid. The questions are examples, or discussion points, and obviously can be altered depending on the subject of the SWOT analysis. Note that many of the SWOT questions are also talking points for other headings—which can be used as you find them helpful, and make up your own to suit the issue being analyzed. It is important to clearly identify the subject of an SWOT analysis, because such an analysis is a perspective of one thing, be it a company, a product, a proposition, and idea, a method, or option, etc. SWOT analysis is commonly presented and developed into a 2 by 2 matrix, which is shown and explained within the SWOT analysis matrix section.

Here the 2 by 2 matrix model automatically suggests actions for issues arising from the SWOT analysis, according to four different categories (Table 10.4):

SWOT analysis came from research conducted at Stanford Research Institute between 1960 and 1970. The background to SWOT stemmed from the need to find out why corporate planning failed. The research was funded by the Fortune 500 companies to find out what could be done about this failure. The research team comprised Marion Dosher, Dr Otis Benepe, Albert Humphrey, Robert Stewart and Birger Lie.

It all began with the corporate planning trend, which seemed to appear first at Du Pont in 1949. By 1960 every Fortune 500 company had a "corporate planning manager" (or its equivalent) and associations of long range corporate planners had sprung up in both the USA and the UK.

However a unanimous opinion developed in all of these companies that corporate planning in the shape of long range planning was not working, did not pay off, and was an expensive investment in futility. It was widely held that managing change and setting realistic objectives which carry the conviction of those responsible was difficult and often resulted in questionable compromises. The fact remained, despite the corporate and long range planners, that the one and only missing link was how to get the management team agreed and committed to a comprehensive set of action programs.

To create this link, starting in 1960, Robert F. Stewart at SRI in Menlo Park California lead a research team to discover what was going wrong with corporate planning, and then to find some sort of solution, or to create a system for enabling management teams to agree and commit to development work, which today we call "managing change." The research carried on from 1960 through 1969. 1100 companies and organizations were interviewed and a 250-item questionnaire was designed and completed by over 5000 executives. Seven key findings lead to the conclusion that in corporations chief executive should be the chief planner and that his immediate functional directors should be the planning team. Dr. Otis Benepe defined the "Chain of Logic" which became the core of system designed to fix the link for obtaining agreement and commitment.

- Values
- Appraise

Table 10.4 **SWOT Analysis**

	Strengths (Internal)	Weaknesses (Internal)
Opportunities (External)	**Strengths/Opportunities** **Obvious Natural Priorities** Likely to produce greatest ROI (return on investment) Likely to be quickest and easiest to implement. Probably justifying immediate action-planning or feasibility study. Executive question: "If we are not already looking at these areas and prioritizing them, then why not?"	**Weaknesses/Opportunities** **Potentially Attractive Options** Likely to produce good returns if capability and implementation are viable. Potentially more exciting and stimulating and rewarding than S/O due to change, challenge, surprise tactics, and benefits from addressing and achieving improvements. Executive questions: "What's actually stopping us doing these things, provided they truly fit strategically and are realistic and substantial?"
Threats (External)	**Strengths/Threats** **Easy to Defend and Counter** Only basic awareness, planning, and implementation required to meet these challenges. Investment in these issues is generally safe and necessary. Executive question: "Are we properly informed and organized to deal with these issues, and are we certain there are no hidden surprises?"—and—"since we are strong here, can any of these threats be turned into opportunities?"	**Weaknesses/Threats** **Potentially High Risk** Assessment of risk crucial. Where risk is low then we must ignore these issues and not be distracted by them. Where risk is high we must assess capability gaps and plan to defend/ avert in very specific controlled ways. Executive question: "Have we accurately assessed the risks of these issues, and where the risks are high do we have specific controlled reliable plans to avoid/ avert/defend?"

- Motivation
- Search
- Select
- Program
- Act
- Monitor and repeat steps 1, 2, and 3

We discovered that we could not change the values of the team nor set the objectives for the team so we started as the first step by asking the appraisal question, for example, what's good and bad about the operation. We began the system by asking what is good and bad about the present and the future. What is good in the present is Satisfactory, good in the future is an Opportunity; bad in the present is a Fault and bad in the future is a Threat. This was called the SOFT analysis.

When this was presented to Urick and Orr in 1964 at the Seminar on Long Range Planning at the Dolder Grand in Zurich, Switzerland they changed the F to a W and called it SWOT analysis.

SWOT was then promoted in Britain by Urick and Orr as an exercise in and of itself. As such it has no benefit. What was necessary was the sorting of the issues into the program planning categories of:

Product (what are we selling?)
Process (how are we selling it?)
Customer (to whom are we selling it?)
Distribution (how does it reach them?)
Finance (what are the prices, costs and investments?)
Administration (and how do we manage all this?)

This is how most organizations manage their businesses which has now been formalized into more safe, efficient and effective means by Operational Excellence.

Assessments and Auditing

Keywords

Accountability; Assessment; Behavior; Checklists; Communicate; Compliance; Controls; Evaluation; Facts; Feedback; Findings; Frequency; Goals; Improvement; Methodology; Noncompliance; Objectives; Parameters; Performance; Plan; Questions; Ranking; Recommendations; Report; Teams; Tracking program.

How to identify what controls are in place to meet requirements, how to verify controls are applied appropriately and achieving required parameters, and identifying noncompliance in proactive constructive manner a for improved performance. Ranking the evaluations and implementing a tracking program for improvements.

OE audits are common technique used to gather sufficient facts and information, including statistical information, typically to verify compliance with baseline requirements. All audits should be viewed as a service to the organization and its workforce and not a burden. OE audits are also different in nature to compliance audits undertaken for governmental regulations. Regulatory compliance audits are typically undertaken by specialized corporate or facility staff or by contracted specialized consulting services. They usually are evaluating an operation or facility for deficiencies as compared to similar arrangements and recommending corrective actions for follow-up. In contrast, OE audits, which can be undertaken by management/supervision serve a different purpose (i.e., continuous improvement), and represent a continuous effort to improve and perfect the implementation and practice of process safety principles during all activities associated with the operation and maintenance of the facility. Management/supervision may conduct these audits formally utilizing agreed processes to develop an action plan. Alternatively, they can conduct audits informally and without creating written reports. Regular and frequent informal audits, along with participation from the work force and supervision/management along with immediate feedback from the supervision/management to the team are considered more effective of the two options.

Applied Operational Excellence for the Oil, Gas, and Process Industries. http://dx.doi.org/10.1016/B978-0-12-802788-2.00011-7
Copyright © 2015 Elsevier Inc. All rights reserved.

Listed below are some areas these audits should include:

• *Presence of basic, safety critical equipment and procedures.* These usually include sensors, transmitters, controller, control valves, pressure reducers, etc. The equipment should have current specification sheets, procedures and maintenance requirements. Employee's knowledge and awareness of safety procedures, duties, rules and emergency response assignments. These documents should be readily available for reference.

• *Regular and effective use of procedures.* Supervision can informally monitor and audit the use and effectiveness of operating procedures. They should ensure procedures are established for all work assignments and they are being followed by operational personnel in all areas of the work assignments. Additionally they need to identify any procedures that need improvement and initiate actions to change and improve them.
• *Compatibility of operating practices with operating procedures.* Sometimes a drift in operating practices may occur, which creates an environment where operating practices may not reflect the actual approved operating procedures. This can be avoided by detecting when it first occurs and also making sure that operating practices are compatible with operating procedures. Employees should be encouraged to identify these circumstances and bring them to the attention of management.
• *Compliance with company standards.* Most company standards and procedures are based on company, industry, or worldwide standards. Supervision should monitor compliance with company standards to ensure effective application and practicality.
• *Adequate Training Program.* Review written training program for adequacy of content, frequency of training, effectiveness of training in terms of its goals and objectives as well as how it fits into meeting the organizational requirements.

Many auditing activities are directed at ensuring quality operator performance, as this has been found to be one of the key factors in preventing incidents. Procedures and safety programs can be audited by formal programs and inspections and tests can be utilized to determine the integrity of the processes and equipment. However, formal audit programs are usually not effective for evaluating human performance or behavior problems. Under close supervision, the line manager, due their daily contact with employees can observe human performance or behavior and assess it to identify concerns more effectively. These observations can also be used to ensure that the quality of training for these individuals is being effectively maintained and that it is also suited for the needs and objectives of the organization. The audits can refer to detailed analysis of incidents to highlight failures or breakdowns that may have occurred and are applicable to their facility, which can be utilized to encourage proper employee actions and behaviors.

Supervisions frequent and regular auditing activities is key in extending accountability for performance, yet at times it may be viewed negatively by employees as a "policing," "enforcement" or "performance evaluation" effort. Supervision should therefore convey the message that the auditing is a training device utilized as part of the process to establish and maintain standards—standards which are in the best interest of the organization as a whole and for the sake of the personal safety of each and every member of the workforce.

In this manner the supervisor is working as a role model and improving the work culture in line with the organizations' goals and objectives.

Audit Checklists

Organizations use internal audits to monitor their compliance with various standards, procedures and regulations. Internal auditors must plan, conduct, and communicate their work effectively. A checklist is an invaluable tool for auditors to use. The checklist should be organized in a meaningful way to quickly assist in planning the audit, performing and documenting the audit findings and reporting the results.

Audit Planning

Audit planning checklists gather information for planning and scoping the audit effectively. Checklists remind the auditor what reports to generate and review to prepare for the audit evaluations. Checklists may also include a questionnaire which the auditor or management completes to give a high-level snapshot of the area under audit review. Audit management relies on planning checklists to ensure that the auditor planned the audit properly.

Audit Fieldwork

During the actual audit, checklists gather information for audit reviews. Certain answers to checklist questions may indicate an audit issue. For example, a "no" answer to the question "Are portable fire extinguishers located at least every 50 feet in the process unit?" is an audit issue if extinguishers are required to be located at those intervals. The checklist should be structured so that negative responses highlight concerns. Checklist questions and answers must be written clearly and in sufficient detail, with references to requirements when appropriate, to support the conclusion.

Audit Review

Checklists are often used to confirm adequate audit work documentation. For example, the question "Do all work papers include a title, reference number, and auditor initials?" indicates whether standards were followed. Additionally, audit management reviews the audit work for quality, consistency and accuracy. For example, the question "Did each test include the required audit sample size?" validates that adequate evidence exists to support the audit results.

Audit Reporting

The audit report is the final product from the review and is communicated to all respective stakeholders. Audit report checklists are developed to ensure relevant information is communicated with the proper tone and in the proper format for maximum impact. These checklists can also remind the auditor who the target audience is and the likely reaction to the audit report. The checklist can also ensure the format for audit reports are in alignment with audit requirements.

The selection of an effective audit team members is critical to the success of the program. Audit individuals should have adequate experience, knowledge and training with auditing techniques, practices and procedures.

The following questions may assist in ensuring a successful audit program is established:

- Is there a written policy on auditing?
- Does the audit policy specify standards in terms of:
    Responsibilities
    Teams
    Locations
    Frequencies
    Methodology
    Follow-up
- Do corporate plans include schedules for auditing?
- Do the corporate plans include:
    Management audits
    Department audits
    Technical audits
    Cross Department audits
    Contractor audits
    Environmental, Safety and Health Audits
    Audit training

- How is the effectiveness of the auditing verified?
- Are audits undertaken at the planned frequencies?
- Are audits undertaken by a wide cross-section of the work force?
- Are individuals from outside the audited organization included in the audit team?
- Are contractor personnel included in any of the audit teams?
- Does management review the findings of the audit teams and undertake follow-up of the audit findings?
- Are audit findings shared or discussed with personnel in the work areas?
- Are lessons from the audits used to improve designs and operations across the organization?
- Is there any treatment made of the audit findings to compare results to other audits?
- How frequently is implementation progress reviewed?
- Are rejections of audit recommendations properly authorized and documented?

Everyone has a role to play in achieving success in operational excellence for an organization. Regardless of the individual role, operator, engineer, manager, office worker—everyone contributes to success in operational excellence. As each individual undertakes their role, everyone is responsible for conducting themselves in accordance with the operational excellence values as expressed by the organization. Establishing a common understanding of each respective role, the company's values, management priorities and aligning worker behaviors and decisions are key to OE success for the organization and important to build up and assure sustainable performance for the long term. Table 11.1 summarizes the major OE players, what is required by them and how this can be achieved, while Figure 11.1 provides sample audit recording forms.

Regardless of how it's done, workplace monitoring is a critical and longstanding function of the safety and internal auditing profession. It is also encouraging to see line managers and selected employees increasingly involved in this function as well. The goal of monitoring is to gain a better understanding of the state of the organization. Leading and lagging indicators also play a role, and are covered in Chapter 8. So what is the best way to monitor the workplace? To answer that question, we need to first look at the three types of active monitoring in common usage.

1. **Inspections**—The National Safety Council (NSC), Accident Prevention Manual defines inspections as activities designed "to locate and repair existing and potential unsafe conditions or activities." In short, inspections are performed to determine if workplace conditions and worker behavior is compliant with ES&H regulations, standards and good practices.
2. **Audits**—These are defined by the National Safety Council (NSC), as "a methodical examination of procedures and practices to verify whether they conform with good ES&H practices auditors base their judgment of compliance or deficiency on the evidence gathered." Audits generally have a larger scope than inspections and may employ large teams of auditors that are often independent (third party) from the area audited.
3. **Assessments**—There are several types of assessment: self-assessment, independent assessment, masnagement assessment, etc., but regardless of the type, all assessments focus on performance rather than compliance and are intended to help drive continuous improvement. Assessments do not ignore unsafe conditions and unsafe acts but their goal is to evaluate the effectiveness of work and safety system performance. Like inspections, assessments are conducted either by teams or individuals.

Table 11.1 OE Team Players, Responsibilities and Actions

Who	What	How
Leaders, leadership team, managers	Build OE culture Demonstrate leadership and commitment Align to OE vision and objectives	Establish vision and objectives Implement OE processes Be accountable Participate in OE and ensure others do Act as a role model Participate in assessments
OE champions	Support OE implementation and operation	Help establish OE processes Coach management and employees in OE Help implementation and governance of OE Assist with OE in business plans Coordinate assessments Undertake OE principles
HSSE teams and SMEs	Assist in establishing standards Provide expertise in area of knowledge Support development, implementation and improvement of OE	Provide HSSE support to OE teams and management Help implementation and governance of OE Help plan and implement OE audits and reports Provide SME for HSSE in OE for improvements Undertake OE principles
OE process sponsors	Provide resources and support to ensure OE success	Serve as advocate for respective OC processes Ensure OE processes are assess periodically for effectively and efficiency Be accountable for the OE processes as assigned Coordinate with other OE processes as required by the business plan
OE process advisors	Provide SME expertise for OE processes	Coordinate and lead OE process efforts Ensure OE processes, documents and records are maintained Coordinate and participate in process measurement activities, reports and improvement activities Ensure contacts and communication maintained with OE entities, SMEs, etc.
All employees and contractors	Operate incident free	Participate in OE processes and recommend improvements Share lessons learned Investigate incidents share causes Undertake OE principles

Summary

An external assessment was conducted of the (Site Name) from (Dates). The assessment team conducted a comprehensive review of the site's higher risk process safety programs, including in-depth reviews to verify the system effectiveness.

(Insert one paragraph summary of findings and suggested improvements)

Besides these improvements, the report documents additional opportunities to further enhance the overall effectiveness of the operating programs. Specific details on the individual systems are provided via a one-page sheet for each system. Additionally, the following are a summary of the highlights, opportunities and scoring.

I. Key System Highlights and Notable Practices
• (Key items that were identified – typically limited to 5)

II. Key Observations
• Assessment was performed consistent with the Industry External Assessment Guidance Documents. Process safety systems reviewed are noted in sections xxx and yyyy of this report.
• Evaluated the quality of the written programs and the effectiveness of field implementation consistent with the approved assessment protocols.
• Regulatory compliance was outside the scope of this assessment.
• Legal review was undertaken by Law Organization.
• (other information as appropriate)

Example Findings Letter

Assessor	Title/Position	System Assessed
		Team Lead
		Process Safety Leadership
		Management of Change
		Mechanical Integrity
		Critical Equipment
		Risk Management
		Safe Work Practices
		Operating Procedures

Audit Team Assignments

Terms	Definitions
Highlights/Good Practices	Identified practices or initiatives that are better than expected. Selected highlights can be reviewed as part of industry benchmarking process for inclusion within a database for sharing.
Observations	Higher priority work practices observed during the assessments that were inconsistent with the assessment guidelines. These should be recorded as statements of fact.
Completed Protocols	Detailed protocols completed by the assessor provide subjects reviewed as well as the general comments regarding the site's process safety programs
Interviews	Job positions of the individuals who were interviewed for each individual system.

Definitions of Terms

Figure 11.1 Sample audit recording forms.

Management System	Scoring		
	System	Effectiveness	Overall
Process Safety Leadership			
Management of Change			
Process Hazard Analysis			
Mechanical Integrity			
Facility Siting			
Safe Work Practices			
Operating Practices			
Management System Average			

Notes:
System score refers to having written plans and programs in place
Effectiveness score refers to the level of execution of the site's process safety programs.
Overall score is a simple average of the two scores.

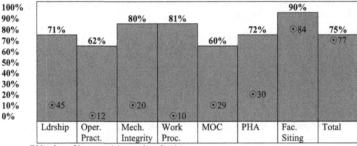

⊙Number of items as 4 (see rating chart)

Performance of Items Weighted as 4

Rating Chart

Level	Description
4	• Excellent* process safety performance is being realized and is supported by advanced, rigorous, robust, integrated and sustainable systems. Might be considered "World Class" performance operation. • Systems Healthy to Healthy and Sustainable** (Sustainable systems must be able to demonstrate a continuous improvement element or elements – the act/adjust portion of the Plan, Do, Check, Act/Adjust cycle (i.e., audits, metrics for performance/effectiveness verification, gap identification, etc.)).
3	• System essentially meets the requirements of the practice. • Desired process or practices are documented and being used to manage the organization. • Systems mostly in place, with only minor gaps observed. • System or practice is tested/ Fixed interval. • Impact of any gaps observed is anticipated to be in low severity and does not affect the meaning or the intent of the item.
2	• Some aspects of the item are addressed, but significant gaps observed. • Impact of any gaps observed is significant enough that it is anticipated to fail to meet the intent of the item. • Systems partially in place, but is not fully implemented, documented or not well understood or practiced. • System or practice is tested on an ad hoc basis, i.e., not at a set interval.
1	• A majority of the aspects of the item are not addressed, with major gaps observed. • Processes or practices not documented or practiced. • Systems are not in place or broadly employed. • Systems are not tested.
0	• Required process, practices or systems do not exist.

*Excellent process safety performance as measured in accordance with API 754, comparative to industry measures of Tier 1 and Tier 2 metrics and the Sites definition excellence relative to Tier 3 and Tier 4 metrics respectively.

**Sustainable systems as defined as being capable of continuing for a considerable period at the same level without being dependent on individuals in a particular position.

Figure 11.1 Cont'd.

System: Process Safety Leadership	System Score:
Assessor:	Effectiveness Score:
	Overall Score:
Highlighted Practices:	
Observations:	
Completed Protocol:	
Interviews:	

Scope	Site Data			Industry Data % Possible			
	Score	Points Possible	%Points Possible	Small Site	Medium Site	Large Site	Average
Leadership Performance							

Weighting	Number of	Total	Possible Points	Percentage
1			5	
2			5	
3			140	
3.5			30	
4			720	
Total			900	

Leadership Summary

Figure 11.1 Cont'd.

Since audits and inspections have similar goals they are combined under the term "inspections" for the purpose of this discussion. It is understood that some organizations use the terms inspections and assessments differently and/or interchangeably but the authors have found the above definitions represent common practice.

Let's first start with a quick example to illustrate the difference between inspections and assessments. Say you had accident where an employee fell from a defective (missing non-slip safety shoe) ladder. You certainly might want to inspect all your ladders for similar and other defects. This is a perfectly reasonable and good thing to do. You might also, however, want to perform an assessment of ladder use safety. Such an assessment would begin by observing workers actually using ladders. From the work observations, employee interviews, and document reviews you would want answers to questions, such as:

- Do workers know the attributes of a safe ladder?
- Do workers check ladders for defects before use?
- Does anyone ever check the ladders?
- Is it acceptable for workers to use defective equipment? If so why?
- Do workers know how to work safely from a ladder?
- Do workers have a means of reporting defective equipment and getting it taken out of service? Do workers ever use it?
- Are safe ladders placed conveniently near the worksite?

The assessment would not stop at merely finding problems but would also look deeper for the causes of problems found.

Inspections—Advantages and Disadvantages

Inspections are a time honored activity in the safety profession and many safety practitioners pride themselves on their encyclopedic knowledge of rules and regulations and their ability to "write up" long lists of deficiencies on their inspections. OSHA's Voluntary Protection Program (VPP) even requires such inspections (monthly for General Industry and weekly for construction) with the entire worksite covered quarterly. Most behavior based safety efforts are also essentially inspections—where predefined behaviors are determined compliant or non-compliant via worker observations. These behavior inspections frequently include immediate coaching and feedback to workers and generally include gathering behavior data that are used in a variety of ways.

The advantages and disadvantages associated with inspections include:

- Inspections are an excellent way to determine if critical equipment is safe to operate, such as checklist reviews of valve alignments, safety equipment availability, etc. Such inspections can often be performed by trained but nonsubject matter experts using well thought-out checklists.
- Inspections generally only identify compliance issues (i.e., unsafe conditions and unsafe acts) which many safety experts view as merely symptoms of deeper and potentially more serious safety system problems. In addition, long lists of noncompliances can be seen as "nit picking" by those inspected.
- Inspections are generally easier to perform, quicker and more straightforward than assessments.
- Inspections are snapshots in time. The time of day, day of week, and work activity at the time of the inspection will determine the findings, which could vary greatly from one point to another.
- Since inspections are so straightforward there is often little room for argument when findings are presented to those responsible.
- Organizations often must spend significant time dealing with corrective actions for issues of little significance. This can act as a diversion from more serious issues important to performance.
- Inspections are a good way to track and trend actions and conditions important to safety. For example, you can do an inspection of desired employee behavior before and after training to see if the training has had the desired effect. Compliance is a low goal and does not lead to continuous improvement. Some view compliance goals as striving to do "just enough to get by."
- Inspections can point out problem areas for further analysis and emphasis.
- You can't inspect in compliance. Compliance inspections tell you little about root causes.

Assessments—Advantages and Disadvantages

Assessments focus on the work and the elements that impact that work, such as strategic planning, worker qualification, procedures, training, staffing, organizational interfaces, communications, etc. A combination of work observations, interviews, and document reviews are common assessment tactics. The goal is to identify root cause problems—and good practices—to promote continuous improvement. Assessments might evaluate a single process or an entire facility and are conducted by teams as well as individuals. Routine self and independent assessment is an expectation for many high reliability organizations, including commercial nuclear power plants and the national defense laboratories. The advantages and disadvantages associated with assessments include:

- Assessments focus on work performance, giving the organization a better understanding of what is actually happening in their facilities. (Recall the managers on the Deepwater Horizon actively inspecting for slip, trip, and fall hazards while completely missing the ongoing safety-critical well capping effort that was rapidly turning into a disaster.).
- Assessments generally take considerably more time than inspections.
- Assessments note noncompliances, negative trends, etc., and use them as prompts to help determine areas where a deeper search for the root causes of those deficiencies is indicated.
- Assessments are harder to do. Assessors must be knowledgeable in the area assessed and trained in the assessment process.
- Assessments generally produce fewer findings for the organization than inspections of similar duration and scope.
- Assessments can be more expensive than inspections.
- Assessments are an excellent way for managers to partner with their employees to seek safer and better ways to perform the work.
- Assessment findings often require considerable assessor judgment making them potentially controversial.

What's the Answer—Inspections or Assessments?

Comprehensive feedback is necessary to truly understand where you are in your OE journey. Inspections can add value to any improvement effort as indicated in the advantages shown above. But traditional compliance inspections that target conditions, and the more recent behavior based observations programs that target behavior compliance, can't be the end of the story. In too many cases we've put too many resources and too much finite safety energy into compliance-based safety inspections and observations, often at the expense of an effective assessment process. We need to look beyond the symptoms and get to the root causes. If not, we are doomed to "Whack-a-Mole" approach to OE where the same problems recur over and over again. Assessments help us make the leap from Whack-a-Mole to a more productive promotion of continuous improvement. They also help us move from a fixation on symptoms and compliance to a deeper understanding of the work and achieving excellent performance. As widely acclaimed safety consultant Dan Petersen stated over 30 years ago in his classic, Techniques of Safety Management, "If we deal only at the symptomatic level we end up removing symptoms and allowing root causes to remain thus leading to another accident.". This is every bit as true today as it was in 1978.

> *Excellence is an art won by training and habituation. We do not act rightly because we have virtue or excellence, but we rather have those because we have acted rightly. We are what we repeatedly do. Excellence, then, is not an act but a habit.*
>
> *Aristotle*

When he said this he was most likely speaking to artists, craftsmen, and students of the day, however Aristotle could very well have been talking to today's oil and gas industry executives when he wrote those words centuries ago. Today's industry executives aspire to the same ambitious goals as Aristotle's listeners. They want their businesses to run safely and more efficiently: to demonstrate operational excellence on a daily basis. But aspirations, intentions, and desires are not enough; excellence is a well-practiced habit that results from millions of small acts, performed by hundreds of thousands of employees every single day.

Acronyms

This section provides a comprehensive listing of acronyms used in this book.

6Ms	Manpower, Methods, Metrics, Machines, Materials, Minutes
ACC	American Chemical Council
AIHA	American Industrial Hygiene Association
AIMS	Asset Integrity Management System
ALARP	As Low As Reasonably Practical
ANSI	American National Standards Institute
API	American Petroleum Institute
ASSE	American Society of Safety Engineers
BP	British Petroleum
CCLA	Corporate Criminal Liability Act
CCPS	Center for Chemical Process Safety
CEO	Chief Executive Officer
CERES	Coalition for Environmentally Responsible Economies
CFR	Code of Federal Regulations
CI	Continuous Improvement
COSHA	China Occupational Safety and Health Administration
CSR	Corporate and Social Responsibility
DCS	Distributed Control System
DMAIC	Define, Measure, Analyze, Improve, and Control
EHS	Environment, Health, and Safety
EMS	Environment Management System
EPR	Extended Product Responsibility
ERM	Enterprise Risk Management
ERP	Emergency Response Plan
ESD	Emergency Shutdown
EU	European Union
HAZOP	Hazard and Operability
HSSE	Health, Safety, Security, and Environment
IFC	International Finance Corporation
ILO	International Labor Organization
JSA	Job Safety Analysis
KPI	Key Performance Indicator
KSAs	Knowledge, Skills, Abilities, and Attitudes
LOPC	Loss Of Primary Containment
LWCRs	Lloyds Weekly Casualty Reports
MOC	Management of Change
MSDS	Material Safety Data Sheet

MVA	Motor Vehicle Accidents
NDE	Net Direct Expenditures
NGO	Non-Governmental Organizations
NSC	National Safety Council
OE	Operational Excellence
OECD	Organization for Economic Co-operation and Development
OEMS	Operational Excellence Management System
OHSMS	Occupational Health and Safety Management System
OI	Operations Integrity
OIMS	Operations Integrity Management System
OMS	Operations Management System
OSHA	Occupational Safety and Health Administration (US)
OSHC	Occupational Safety and Health Center
PDCA	Plan, Do, Check, Act
PHA	Process Hazard Analysis
PPE	Personal Protective Equipment
PPORD	Product and Process-Oriented Research and Development
PRD	Pressure Relief Device
PrHA	Preliminary Hazard Analysis
PSE	Process Safety Event
PSM	Process Safety Management
PSSR	Pre-Startup Safety Review
QA/QC	Quality Assurance/Quality Control
QRA	Quantitative Risk Assessment
RCA	Root Cause Analysis
REACH	Registration, Evaluation, and Authorization of Chemicals
RMP	Risk Management Plan
SHE	Safety, Health, and Environment
SME	Subject Matter Expert
SMS	Safety Management System
SP	Social Performance
SWOP	Strengths, Weakness, Opportunities, and Threats
TQM	Total Quality Management
VGCL	Vietnam General Confederation of Labor
VPP	Voluntary Protection Program

Definitions

This section explains all the unique terms utilized in this book.

Accident See Incident.

Best Practice A suggested approach or method, adopted, or proven in the industry, to effectively achieve superior results for a desired objective.

Cause An event, situation, or condition which results, or could result, directly or indirectly to an incident.

Consequence The direct undesirable result of an accident sequence usually involving a fire, explosion, release of toxic material. Consequence descriptions may include estimates of the effects of an accident in terms of factors such as health impacts, physical destruction, environmental damage, business interruption, and public reaction or company prestige.

Continuous Improvement An ongoing effort to improve products, services, or processes. These efforts can seek "incremental" improvement over time or "breakthrough" improvement all at once.

Contributing Causes Factors that, by themselves, do not lead to the conditions that ultimately caused the event; however, these factors facilitate the occurrence of the event or increase its severity. Sometimes it is also referred to as causal factors or immediate causes.

Effectiveness The capability of producing a desired result. When something is deemed effective, it means it has an intended or expected outcome.

Efficiency The extent to which time, effort, or cost is well-used for the intended task or function. It often comprises specifically the capability of a specific application of effort to produce a specific outcome effectively with a minimum amount or quantity of waste, expense, or unnecessary effort.

Emergency Response Actions to mitigate the consequences of an emergency, for the protection of human health and safety, quality of life, property, and the environment. It may also provide a basis for the resumption of normal social and economic activity.

Excellence A talent or quality which is unusually good and so surpasses ordinary standards. It is also used as a standard of performance.

Expectation Specific requirements for an organization, individual or transaction that define the consistent and systematic approach for achieving pre-defined organizational outcomes in alignment with industry leading performances the sum of which describes the character industry leading performance, i.e., Operational Excellence.

Gap Analysis The comparison of actual performance with potential or desired performance. If a company or organization does not make the best use of current resources, or forgoes investment in capital or technology, it may produce or perform below its potential.

Hazard Source of or situation that could result in harm or adverse consequences in terms of injury or illness, damage to the environment, property, or the workplace, business interruption, impact on the reputation of the organization, or a combination of these.

Hazard Control   That function in an organization directed toward the recognition, evaluation, and reduction or elimination of the destructive effects of hazards emanating from human acts of commission and omission and from the physical and environmental aspects of the workplace.

Implementation   The systematic and structure process of converting expectations into specific actions and instilling behaviors and attitudes (i.e., culture) required for a successful and sustainable change at all levels of an organization.

Incident   An event or sequence of events that results in undesirable consequences.

Injury   Physical harm or damage to a person resulting from traumatic contact between the body and an outside agency or exposure to environmental factors.

Job Safety Analysis (JSA)   A safety management tool that is utilized to define and control the hazards associated with a process, job, task, or procedure. Each step of the undertaking is analyzed for hazards and safeguards to ensure they are as low as reasonably practical.

Key Performance Indicator (KPI)   Measurement statistics utilized to evaluate the performance of an organization. KPIs are usually categorized as leading or lagging. Lagging safety indictors are commonly incident statistics, such as injuries, fatalities, MVAs, fires, and are considered lagging due to that they materialize after the incidents. Leading safety indicators are usually number of management safety inspections, number of safety meetings, levels of safety training, etc. and are considered proactive safety activities to prevent an incident.

Lessons Learned   Identification of root cause failures from previous incidents to provide improvements to ongoing operations or design to prevent future incidents from occurring from these incidents.

Loss Prevention   A functional effort to identify and correct potential incident concerns, before they result in an actual event resulting in a loss.

Management of Change (MOC)   A process to evaluate and manage variations to a facilities' operations and equipment to ensure that potential risks arising from these changes remain at an acceptable level.

Nonconformance   Any deviation from planned activities within a management system.

Operational Excellence (OE)   Operational Excellence (OE) is an element of organizational leadership that stresses the application of a variety of principles, systems, and tools toward the sustainable improvement and industry leading performance of the organization, which can be measured by key performance indicators. The process involves focusing on customer needs, keeping employees positive and empowered and continuous improvement of activities in the workplace, while adhering to the highest standards for Safety, Health, and Environmental Stewardship in a Cost-Effective and Profitable manner.

Operational Excellence System   The synergistic compilation of programs, processes, practices, methods, tools, technologies, and organizational capabilities by which organizations consistently and systematically achieve desired OE performance and world class results.

Pre-Startup Safety Review (PSSR)   A final check, initiated by a trigger event, prior to the use or reuse of a new or changed aspect of a process. It is also the term for the OSHA PSM and EPA RMP element that defines a management system for ensuring that new or modified processes are ready for startup.

Process   Any management function, activity, or operation performed methodically and systematically leading to a particular event or outcome. Processes represent the means for transforming expectations into specific actions to deliver the desired change.

Process Safety   A disciplined framework, for managing the integrity of operating systems and processes by applying good design principles, appropriate engineering, and applicable operating and maintenance practices.

Process Safety Management (PSM) Comprehensive set of plans, policies, procedures, practices, administrative, engineering, and operating controls designed to ensure that barriers to major incidents are in place, in use, and are effective.

Process Safety Review An inspection of a plant or process unit, drawings, procedures, emergency plans, and/or management systems, etc. usually by an onsite team and usually problem solving in nature.

Product Stewardship The fashion in which an organization who designs, makes, sells, uses, and recycles, or disposes of their products can help protect safety, health, and environmental quality.

Qualitative Methods A type of evaluation that is gained through experience, e.g., application, operational support, or essential, required, and desirable features.

Quality A characteristic, innate or acquired, that, in some particular, determines the nature and behavior of a person or thing.

Risk The combination of expected likelihood or probability (e.g., events/year) and consequences or severity (effects/event) of an incident, i.e., $R = f\{P, C\}$.

Risk Analysis The development of a qualitative and/or quantitative estimate of risk-based on engineering evaluation and mathematical techniques (quantitative only) for combining estimates of event consequences and frequencies.

Risk Criteria Terms of reference against which the significance of a risk is evaluated. Risk criteria is based on organizational objectives, derived from external and internal context. Commonly determined by the need to comply with standards, laws, policies, and objectives of the organization.

Risk Evaluation The process of comparing the outcomes of risk analysis with risk criteria to determine the risk magnitude against the organization's risk acceptability criteria or tolerance.

Risk Management The systematic application of policies, programs, procedures, and practices to the tasks of analyzing, assessing, and controlling risk in order to protect employees, the public, and the environment as well as company assets, avoidance of business interruptions. It includes decisions to use suitable engineering and administrative controls for reducing risk.

Risk Tolerance Degree to which an entity, asset, system, network, or geographic area is willing to accept risk.

Root Cause The most basic causes of an incident that can reasonably be identified which management has control to fix and for which effective recommendations for preventing reoccurrence can be generated. Sometimes it is also referred to as the absence, neglect, or deficiencies of management systems that allow the "causal factors" to occur or exist such as failure of particular management systems that allow faulty design, inadequate training, or deficiencies in maintenance to exist. These, in turn, lead to unsafe acts or conditions which can result in an accident.

Safety A general term denoting an acceptable level of risk of, relative freedom from, and low probability of harm.

Safety Culture The collective individual and group values, attitudes, competencies and patterns of behavior that determine the commitment to and style and proficiency of an organization's health and safety programs.

Safety Dashboard Typically an intranet webpage of an organization which features various safety statistics (e.g., leading and lagging indicators). These are arranged in a dial or graph format, in which real-time statistics (i.e., weekly, monthly reports) are compared to stated safety targets or goals.

Safety Management System (SMS)   A set of related elements that establish or support loss prevention policies, programs, objectives and mechanisms to achieve those goals in order to maintain and improve an organizations safety culture and achieve high levels of safety performance. See also Process Safety Management.

Stakeholders   Person or persons impacted or potentially impacted by the organization's operations. These may include employees, stockholders, neighbors, emergency responders, other industries, competitors, commercial partners, public at large, regulators, and other entities with a personal stake in the organization's operations.

System   A collection of people, machines, and methods organized to accomplish a set of specific functions.

Bibliography

[1] ANSI/AIHA, Occupational Health and Safety Management Systems (OHSMS), Z10-2012, ANSI/AIHA, Fairfax, VA, 2012.

[2] American Petroleum Institute (API), RP 75, Development of a Safety and Environmental Management Program for Offshore Operations and Facilities, API, Washington, DC, 2004.

[3] American Petroleum Institute (API), RP 754, Process Safety Performance Indicators for the Refining and Petrochemical Industries, first ed., API, Washington, DC, 2010.

[4] W. Booth, S. Lindborg, Handbook to Achieve Operational Excellence, Reliabilityweb.com Press (Online Publisher), 2011.

[5] I. Cameron, R. Raman, Process Systems Risk Management, Process Systems Engineering, vol. 6, Elsevier Academic Press, San Diego, CA, 2005.

[6] Center for Chemical Process Safety (CCPS), Guidelines for Implementing Process Safety Management Systems, second ed., American Institute of Chemical Engineers, New York, NY, 2015.

[7] Center for Chemical Process Safety (CCPS), Conduct of Operations and Operational Discipline: For Improving Process Safety in Industry, CCPS/Wiley, New York, NY, 2011.

[8] Center for Chemical Process Safety (CCPS), Guidelines for Safe Process Operations and Maintenance, American Institute of Chemical Engineers, New York, NY, 1995.

[9] Center for Chemical Process Safety (CCPS), Guidelines for Risk Based Process Safety, American Institute of Chemical Engineers, New York, NY, 2007.

[10] Center for Chemical Process Safety (CCPS), Guidelines for Process Safety Metrics, American Institute of Chemical Engineers, New York, NY, 2009.

[11] Center for Chemical Process Safety (CCPS), Process Safety Leading and Lagging Metrics, American Institute of Chemical Engineers, New York, NY, 2008.

[12] Center for Chemical Process Safety (CCPS), The Business Case for Process Safety, second ed., American Institute of Chemical Engineers, New York, NY, 2006.

[13] K.J. Duggan, Design for Operational Excellence: A Breakthrough Strategy for Business Growth, McGraw-Hill, New York, NY, 2011.

[14] European Process Safety Center (EPSC), Making the Case for Leading Indicators in Process Safety, 2012.

[15] K. Glac, The Influence of Shareholders on Corporate Social Responsibility, History of Corporate Responsibility Project Working Paper No. 2, Center for Ethical Business Cultures, Minneapolis, MN, 2010. p. 15.

[16] J.M. Haight (Ed.), Handbook of Loss Prevention Engineering, Wiley-VCH, Weinhwim, Germany, 2013.

[17] Health and Safety Executive (HSE), Step by Step Guide to Developing Process Safety Performance Indicators, HSG 254, Sudbury, Suffolk, UK, 2006.

[18] International Standards Organization (ISO), ISO 31000, Risk Management – Principles and Guidelines, ISO, Geneva, Switzerland, 2009.

[19] C.J. Fombrum, Leading Corporate Change, McGraw Hill, New York, NY, 1994.

[20] F.A. Manual, Leading and lagging indicators: do they add value to the practice of safety? Professional Safety 54 (12) (December 2009) 42–47.

[21] J.S. Mitchel, Operational Excellence: Journey to Creating Sustainable Value, Wiley, 2015.

[22] National Safety Council (NSC) Accident Prevention Manual, Accident Prevention Manual, twelfth ed., NSC, Itasca, IL, 2000.

[23] National Safety Council (NSC), in: G. Swartz (Ed.), Safety Culture and Effective Safety Management, NSC, Itasca, IL, 2000.

[24] National Safety Council (NSC), Initiating a Safety Management System, NSC, Itasca, IL, 2000.

[25] C. Nilson, Games that Drive Change, McGraw Hill, New York, NY, 1995.

[26] D.P. Nolan, Handbook of Fire and Explosion Protection Principles for Oil, Gas, Chemical and Related Facilities, third ed., William Andrew/Elsevier Publications, New York, NY, 2014.

[27] D.P. Nolan, Economic Justification of Safety Management Systems, Safety in Focus, Saudi Aramco, Dhahran, Saudi Arabia, April 2015.

[28] Occupational Safety and Health Administration (OHSA), Voluntary safety and health program management guidelines, Federal Register 54 (16) (1989) 3904–3916, US Government Printing Office, Washington, DC.

[29] D. Peterson, Techniques of Safety Management, a Systems Approach, third ed., Aloray Pub, Goshen, NY, 1989.

[30] D. Pope, Process: Understand the Elements of (PSM) 29CFR 1910.119, CreateSpace Independent Publishing/Amazon (Web Online), 2012.

[31] C. Price-Kuehne (Ed.), The 100 Largest Losses 1972–2011, Large Property Damage Losses in the Hydrocarbon Industries, twenty-second ed., Marsh Global Risk Engineering, London, UK, 2012.

[32] S.O. Shakioye, J.M. Haight, Modeling using dynamic variables: an approach for the design of loss prevention programs, Safety Sciences 48 (1) (April 2009) 46–53.

[33] I. Sutton, Offshore Safety Management, Implementing a SEMS Program, second ed., William Andrew/Elsevier, Waltham, MA, 2014.

[34] I. Sutton, Process Risk and Reliability Management, Operational Integrity Management, second ed., Gulf Professional Publishing/Elsevier, NY, New York, 2014.

[35] K. Roose, Nuns Who Won't Stop Nudging, The New York Times, Business Day, November 11, 2011.

[36] A. Sullivan, S.M. Sheffrin, Economics: Principles in Action, Pearson Prentice Hall, Upper Saddle River, New Jersey, 2003.

[37] UK Health and Safety Executive (HSE) HSG-254, Developing Process Safety Performance Indicators, HSE, London, UK, 2006.

Index

Note: Page numbers followed by "f" and "t" indicates figures and tables respectively.

Printed in the United States
By Bookmasters